Edward Jarvis

Primary Physiology

Edward Jarvis

Primary Physiology

ISBN/EAN: 9783743318144

Manufactured in Europe, USA, Canada, Australia, Japa

Cover: Foto ©berggeist007 / pixelio.de

Manufactured and distributed by brebook publishing software (www.brebook.com)

Edward Jarvis

Primary Physiology

PRIMARY PHYSIOLOGY,

FOR SCHOOLS.

By EDWARD JARVIS, M. D.,

AUTHOR OF

"PHYSIOLOGY AND LAWS OF HEALTH."

NEW YORK:
A. S. BARNES & Co., 111 & 113 WILLIAM STREET,
(CORNER OF JOHN STREET.)

1866.

PREFACE.

The following work is intended to explain those general principles of Anatomy and Physiology which every one should understand in order to manage his own body successfully. These principles should be taught, illustrated, and enforced in every school; and no boy or girl should go forth from the school to the world without this preparation for the responsibilities which must come upon every one for his self-management.

In a small work like this, it is impossible to explain more than the leading elementary principles of Anatomy and Physiology, and their most obvious applications to the ordinary exposures and habits of life.

But for those who have sufficient interest and time to pursue the subject further, in school or elsewhere, I have prepared a larger work, the "PHYSIOLOGY AND LAWS OF HEALTH," in which these principles are explained more minutely, and their applications to the duties and the liabilities of life illustrated to a much greater extent, than can be done here.

EDWARD JARVIS.

DORCHESTER, MASS.,
October, 1865.

CONTENTS.

INTRODUCTION.
Chap. Page.
I. Anatomy and Physiology, Uses and Objects of,..............7

FOOD AND DIGESTION.
II. Food. Mouth. Teeth,..................................9
III. Saliva. Mastication. Œsophagus. Swallowing,............12
IV. Stomach, — Coats,......................................15
V. Gastric Juice. Hunger, ...19
VI. Digestive Action. Chyme. Pyloric Valve,.................22
VII. Time and Conditions of Digestion. Beaumont's Observations. Effect of Drink with Food, and of Mastication and Texture of Food on Digestion,..........................26
VIII. Alimentary Canal. Lacteal Absorbents and Ducts. Chyle,..30
IX. Conditions of Eating. Appetite. Quantity of Food. Rest before and after Eating,..................................34
X. Quality of Food. Condiments and Stimulants. Frequency of Meals. Breakfast. Dinner. Supper,...................38

CIRCULATION OF THE BLOOD.
XI. Apparatus of Circulation. Heart. Valves,...............42
XII. Arteries and Veins. General and pulmonary Circulation. Action of the Heart. Quantity of Blood in the Body,......45

NUTRITION.
XIII. Growth. Changes of Particles. Nutrition. Absorption. Dead Atoms,...52

RESPIRATION.
XIV. Venous Blood. Situation of Lungs. Chest,...............57
XV. Chest. Muscles. Expansion and Contraction of Diaphragm,..62
XVI. Lungs. Windpipe. Vocal Chords. Air-Tubes,..............66
XVII. Air and Blood-vessels of Lungs. Coughing, ...70
XVIII. Inspiration. Expiration. Dead Atoms. Air,..............74

Chap.		Page.
XIX.	Purification of the Blood. Corruption of the Air. Carbonic Acid and Water carried off. Foul Air of crowded Rooms,	73
XX.	Quantity of Waste carried out through the Lungs. Size, Shape, and Motions of the Chest,	82
XXI.	Air corrupted. Fresh Air needed. Ventilation,	87
XXII.	ANIMAL HEAT.	89

SKIN.

XXIII.	Cuticle. Effect of Friction. Blisters. Corns. Seat of Color,	92
XXIV.	True Skin. Perspiration. Oily Excretions. Temperature of Rooms,	96
XXV.	Cutaneous Absorption. Sensibility. Touch,	101
XXVI.	Clothing, — Quantity, Materials. Unclean Garments, Airing of,	104
XXVII.	Foul Skin. Bathing. Cold Bath, good Effects of,	108

BONES.

XXVIII.	Bones, Composition, Nutrition, Sensibility of. Skeleton,	112
XXIX.	Skull. Back-bone. Chest. Wrist. Hand. Foot. Strength of Frame,	116
XXX.	Joints. Synovial Membrane. Dislocations. Sprains,	121
XXXI.	Attitude, erect, stooping. Curved Spine,	124

MUSCLES.

XXXII.	Muscles, Character, Situation; of Elbow; of Fingers; Form; Action, voluntary, involuntary,	127
XXXIII.	Muscles, controlled by the Will, coöperate; grow by Use. Exercise, Effect of, on Health; Amount; Kinds; Condition of,	132
XXXIV.	Labor, Limit to; Excess of; violent; in Youth. Condition of Labor. Day and night Labor. Effect of Spirit and Cheerfulness on Labor,	137

BRAIN.

XXXV	Brain. Spinal Cord. Nerves,	142
XXXVI.	Nerves of Sensation and Motion. Connection of Body and Brain,	145
XXXVII.	Brain and Mind. Connection of Mind and Body,	149
XXXVIII.	Eye,	152
CONCLUSION,		156

PRIMARY PHYSIOLOGY.

INTRODUCTION.

CHAPTER I.

USES OF ANATOMY AND PHYSIOLOGY.

1. *Anatomy* is the description of the various parts and organs of the animal body.

2. *Physiology* is the description of the uses of these organs, and of the manner in which they operate.

3. A knowledge of the anatomy and physiology of the human body is necessary, in order to understand the uses of our organs, the extent of their powers, and the purposes to which they may be applied.

4. It is necessary to have this knowledge, in order to govern the body correctly, and to apply its powers to their proper purposes, and to prevent its suffering any injury or derangement.

5. Every one is appointed to take charge of his own body, to supply its wants, to govern its appetites, to direct its motions, and to aid in its purification.

6. Every one must supply himself with food of

the proper quantity and quality, which the stomach can digest, and which will nourish the body.

7. He must supply his lungs with pure air to breathe, in sufficient quantity to purify his blood.

8. He must cleanse his skin from all impurities, and protect it from cold by clothing, shelter, and fire.

9. He must exercise his muscles sufficiently to keep his body in good health, and he must so arrange and govern his labor, that his health shall not suffer from excessive exertion.

10. He must exercise his mind sufficiently for the life and health of the brain and nervous system; but he must restrain his mental action, so that the brain and nervous system shall not suffer from overexertion.

11. He must govern all his passions and appetites, so that they may always subserve the purposes of life.

12. In order to fulfil these several responsibilities, it is necessary for every one to understand the anatomy and the physiology of the organs of digestion, respiration, and of the circulation of the blood, and also of the skin, the bones and muscles, and the brain and nervous system.

13. It is also necessary to know the nature of food and its relation to the digestive organs and the body, of air and its relation to the blood, the effect of bathing and clothing, of exercise and labor, and the effects of the operations of the mind upon the brain and the whole body.

FOOD AND DIGESTION.

CHAPTER II.

FOOD. — MOUTH. — TEETH.

Food.

1. Food is an article of apparently the first and most constant necessity in all animals. None can be long deprived of it, without suffering pain, losing strength, and wasting in flesh.

2. The food is first received into, and ground in, the mouth; next it is digested or changed into pulp in the stomach; then it is carried to the blood-vessels, and converted into blood; and lastly it is carried all over the body, and changed into flesh.

Objects of Food.

3. *Growth.* — By means of the food, constant additions are made to all parts of the body, and thus it grows from the infant's size to the fulness of stature. This indicates the necessity of frequent supplies of food during the period of growth.

4. *Change of Atoms.* — But when the body ceases to grow, the necessity of food does not cease. It is needed as long as life continues, because the body is constantly undergoing a waste or loss of its substance, through the lungs, skin, &c. This incessant waste of material would soon reduce the size and weight of the body, if it were not supplied by new additions from food.

5. The living animal body is composed of materials that were originally dead food. This dead food is very different from the living flesh; and yet, by means of the digestive organs and the digestive and nutritive processes, established by the Creator, the lifeless bread is made flesh, and becomes a part of the living body.

Digestive Organs.

6. The digestive organs consist of the mouth, the gullet, the stomach, the intestinal canal, &c.

7. *The mouth* includes the lips, the cheeks, the tongue, the teeth, and the salivary glands.

Teeth.

8. There are thirty-two teeth in the human mouth. Of these, there are, in each jaw, four

front or *incisor teeth*, (Fig. I. 1, 1,) which are thin and sharp, for the purpose of cutting food; two eye and stomach or *cuspid teeth*, with pointed tops and very long fangs or roots, (Fig. I. 2;) two *bicuspid teeth*, (Fig. I. 3, 3.) These are the same in each jaw; behind these are six grinders or *molar teeth*, with broad tops and uneven surfaces, for the purpose of grinding food, and with three roots in the upper jaw, (Fig. I. 4, 4, 4,) and two roots in the lower jaw. (Fig. I. 5, 5, 5.)

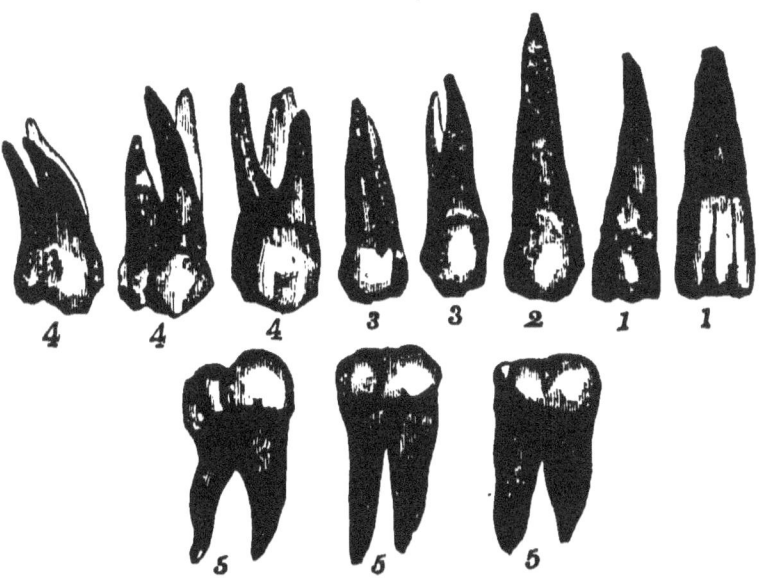

Fig. I.— *Teeth of one Side.*

9. The teeth are composed of a somewhat soft bone within, but they are covered on the

outside with a very dense substance, called the *enamel.* This covering admits of a very fine polish, and is sufficiently hard to resist the effect of friction, so that the teeth are seldom worn.

10. If the teeth are thoroughly cleansed, and the particles of food never allowed to remain on them, they will rarely decay; but if they are neglected, even the enamel will decay; and when this covering is eaten through, and the inner substance of the tooth comes in contact with the food, liquid, and air, the decay goes on rapidly.

11. The inner portion of the teeth is supplied with a nerve, and when this is exposed to food and air in consequence of the decay of the tooth, it suffers very acute pain.

CHAPTER III.

SALIVA.— MASTICATION.— SWALLOWING.

SALIVA.

1. IN good health, the mouth is never dry. It is kept moist by means of several little fleshy bodies, called *salivary glands,* which are placed

in the cheeks and under the tongue. The saliva, or fluid of the mouth, is prepared in these, and poured out from them into the mouth, as the perspiration is poured from the skin upon the forehead.

2. These glands are more active, and pour out more saliva, when the mouth is in motion, especially when we are eating, than when it is still. The flow of this fluid can be increased by chewing any substance, or by moving the cheeks upon the teeth, or by rubbing the point of the tongue upon the glands beneath it.

3. The sole object of the saliva is to moisten the mouth when it is still, and to wet the food when it is eaten. To excite the glands, and to increase the flow of this fluid for any other purpose, by chewing tobacco, or any matter excepting food, or by the offensive habit of spitting, is unnatural as well as disgusting.

Mastication.

4. Each of the works of nature is fitted for a definite purpose, and for no other. The mouth is fitted to masticate or to grind and soften the food preparatory to its digestion in the stomach. The teeth can bite and grind all the proper kinds of food, but nothing else,

without danger of breaking or of decay. The salivary glands can supply all the liquid necessary to soften this food, but no more with perfect safety.

5. Mastication is the first and necessary step in the work of digestion. Before the food can be dissolved in the fluids of the stomach, it must be ground by the teeth, softened by the saliva, and reduced to a pulp in the mouth. The teeth, if in good health, are sufficient for the grinding, and the glands will pour out fluid enough for the softening. No other liquid and usually no drink need be taken while we are eating.

Œsophagus.

6. The *œsophagus*, or gullet, is a soft and fleshy tube, that connects the mouth with the stomach; it opens, at its upper end, into the back part of the mouth, and, at its lower extremity, into the left end of the stomach. It runs along the side of the back-bone, through the neck and chest, behind the windpipe and the great air-vessels.

7. This tube is composed, in part, of muscular fibres, that surround it like rings, following one after another from the top to the bottom

These muscular fibres can be stretched out to enlarge the size of the tube and admit the food or liquid, and they have the power of contraction and of diminishing the channel when nothing is in it.

Swallowing.

8. When the food is sufficiently masticated, it is carried by the tongue to the back part of the mouth, and thrown into the *pharynx*, or the upper part of the œsophagus. When the food or any other substance is received into this tube, the rings of muscular fibres above it contract, and press it downward. Then the next ring contracts and carries the matter farther onward; and thus they all contract, one after another, and carry the food through the tube from the mouth to the stomach.

CHAPTER IV.

STOMACH.

1. THE stomach is an oblong, soft, and fleshy bag, placed in the upper part of the abdomen,

just below, and partially behind, the short ribs, and extends across from the right to the left side.

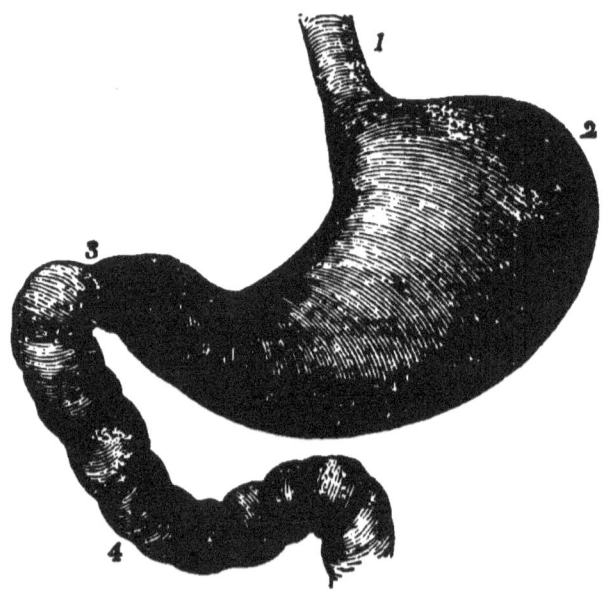

Fig. II.— *Stomach.*

1. Œsophagus.
2. Left end.
3. Pyloric valve.
4. Duodenum, the beginning of the Intestinal Canal.

2. The size of the stomach differs with varying circumstances. Sometimes, when it is distended after a full meal, or with a great quantity of drink, it may contain as much as two quarts. At other times, when food and drink have passed out of it, it may be contracted so as to hold less than one pint. The average size, in an adult of moderate habits of eating, is about that of a quart measure.

Coats of the Stomach.

3. The stomach is composed of three coats or layers of flesh, each of which performs a separate and independent part in the work of digestion, and yet all coöperate to effect the same end. These coats are the outer or *peritoneal*, the middle or *muscular*, and the inner or *mucous* coat.

4. The *peritoneal coat* covers all the outside of the stomach, and gives it strength and support. It also covers the other digestive organs, and is attached to the back-bone, and thus sustains all the abdominal organs in their respective places.

5. The *muscular coat* is composed of stringy fibres, such as compose the lean meat which we see on our tables. These fibres have an active power of contracting or drawing themselves up, and thus diminishing their length, like the earth worm; and they can be stretched out loosely. But they have no active power of expansion.

6. Some of these fibres wind around the stomach in the form of rings, and are placed so closely together as to form an entire, and, in some parts, a very thick coat. Other fibres run

lengthwise of the organ, and thus a double muscular coat is formed.

7. When these circular fibres contract, they reduce the size of the sack, and press upon its contents; and when they contract successively from end to end of the stomach, like similar fibres in the gullet, they press its contents onward. When the longitudinal fibres contract, they shorten the length of the stomach, and press its contents toward the lower end.

8. The *mucous coat* lines the inside of the stomach. This is a thick, soft, and loose membrane. It is not elastic, like the other coats; but it is drawn into wrinkles or folds, when the sack is contracted, and spread out more smoothly, when it is expanded. This coat prepares a mucous or slimy matter, that protects the stomach from the irritation of its contents, and also a fluid in which the food is dissolved.

9. The structure of these three coats is familiarly shown by examining a piece of tripe, which is a preparation of the cow's stomach. The soft, spongy, and porous layer on one side, is the mucous coat; the thick, tough, and fatty layer on the opposite side, is the peritoneal coat; and the layer of lean meat, consisting of stringy fibres in the middle, is the muscular

coat. The coats of the human stomach are arranged in a similar manner.

10. These three coats perform separate offices. The outer gives support to the stomach, the middle gives it motion, and the inner coat prepares the *gastric juice*, or the fluid in which the food is dissolved and digested.

CHAPTER V.

GASTRIC JUICE.—HUNGER.

Gastric Juice.

1. The gastric juice is prepared in, and poured out of, the mucous coat, or inner lining of the stomach, as the sweat is prepared in, and poured out of, the skin. This is a very powerful fluid, and capable of dissolving all the proper food, of every kind, whether of animal or vegetable origin. It is peculiar to the living stomach. Nothing like it can be found out of the living animal, or can be prepared by human skill.

2. *Flow of the Gastric Juice.* — The gastric juice is necessary for the work of digestion, but

it is not always present in the stomach. It is only when some food or other matter is there, that the lining, or mucous membrane, is stimulated to prepare and throw it out. Nor even then does it flow out with a gush, or in any great quantity at once; but it oozes out slowly, drop by drop, as the saliva flows from the glands in the mouth, during mastication.

3. When a morsel of food is in the mouth, the salivary glands are excited, and pour out their fluid until it is completely moistened. When this morsel is swallowed, it excites the mucous membrane of the stomach to pour out its gastric juice, which continues to flow until all the particles of the food are completely mixed and wet with it; and then, when this is done, the stomach is ready to receive and moisten another morsel.

4. *Limit of Gastric Juice.* — Dr. Beaumont says, that the quantity which can be prepared at one time corresponds to the quantity of food which the body then needs for its nourishment. The lining membrane of the stomach begins to pour out this fluid, as soon as one morsel of food is swallowed, and continues to pour it out as long as fresh morsels are brought there, or until enough is provided to digest as much food

as the body then wants for its nutrition. When this limit is reached, no more gastric juice will be given out.

5. *Excess of Food.* — The healthy stomach has power to digest as much food as its gastric juice mixes with in due proportion. But if more food is eaten than the gastric juice can moisten, the excess will remain undigested in the organ; or, being mixed with that which has been previously swallowed, it retards the digestion of the whole.

HUNGER.

6. When the body is in want of nourishment, and the stomach is ready to prepare and pour out gastric juice to dissolve it, the sensation, which we call *hunger*, is felt in the stomach.

7. True hunger begins when the frame wants more nourishment, and the stomach can digest food, and continues as long as the stomach gives out gastric juice; and when this fluid ceases to flow, hunger ceases to be felt. On the contrary, when this sensation ceases to be felt, it is a sign that no more of the digesting fluid can be given out.

8. When a man swallows his food no faster than the gastric juice is prepared to be mixed

with it, and stops as soon as the hunger ceases, he eats no more than he can easily digest; and if he notices the cessation of hunger, he finds a guide to the quantity which he should eat.

CHAPTER VI.

DIGESTIVE ACTION.

1. When the meal is all eaten, the muscular coat of the stomach continues its action, by first contracting and pressing upon the food, and then relaxing and allowing it to lie more loosely.

2. All the fibres of this coat do not contract together, but first one and then another acts, and thus the food is pressed from side to side and forward and backward; and by these varied actions of the muscular fibres, a sort of churning of the stomach upon its contents is kept up, and the food is kept in motion.

3. The agitation of the food is aided by the action of the respiratory organs. The digestive organs (Fig. III. 7, 8) lie between the diaphragm (Fig. III. 3, 3) above and the abdominal muscles below. (Fig. III. 9.) The *dia-*

DIGESTIVE ACTION. 23

phragm (Fig. III. 3, 3) is a great muscle which lies across the bottom of the chest, and moves alternately upward and downward.

Fig. III. — *Relative Situation of the Contents of the Chest and Abdomen.*

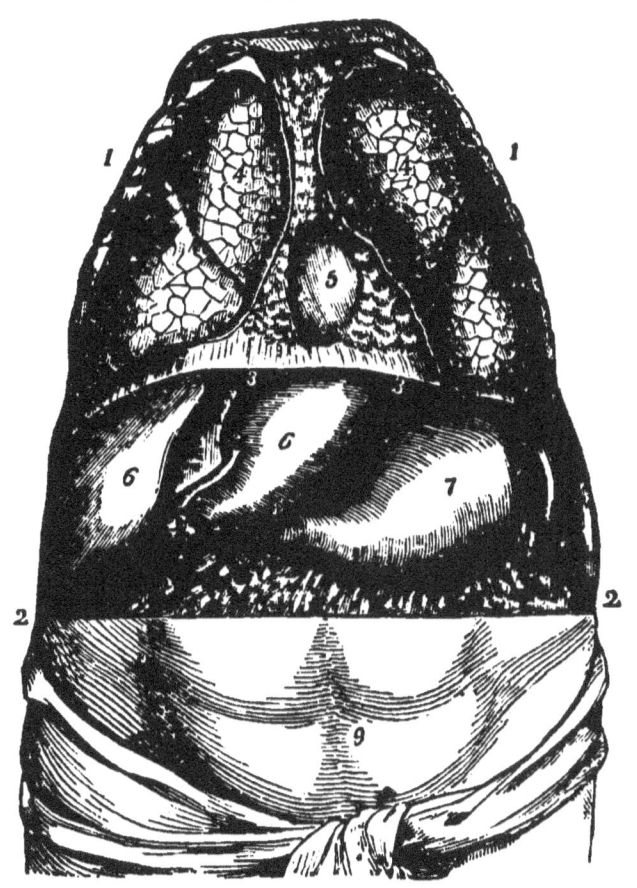

1, 1. Chest. 4, 4. Lungs. 7. Stomach.
2, 2. Abdomen. 5. Heart. 8. Intestinal canal, &c.
3, 3. Diaphragm. 6, 6. Liver. 9. Abdominal muscles.

4. When we take in our breath, the diaphragm

is drawn down and presses upon the stomach, which is next below it. And when we expire, the abdominal muscles — or the front walls of the abdomen — press upon the stomach and the diaphragm, and thus force the air out of the lungs.

5. Every time we breathe, we press the stomach downward by drawing in the air, and upward by expelling it; and thus the act of respiration keeps the stomach and its contents in incessant motion.

CHYME.

6. By this agitation of the stomach, the food and the gastric juice become completely incorporated together. In this mixture all the original differences of the various kinds of food are lost. No trace can there be found of the meat, bread, and vegetables that have been eaten. All are reduced to one homogeneous pulp, called *chyme.*

7. As fast as any portion of the food is digested, it is sent onward, through the right end of the stomach, into the alimentary canal. But, although the digested portions of the food are allowed to pass out, the undigested portions are retained in the stomach by a valve or doorkeeper provided for the purpose.

Pyloric Valve.

8. The stomach is large at its left and small at its right end. (Fig. II. 2, 3, page 16.) The circular fibres of the muscular coat, which covers it all over, are gathered at the right end, so as to form a thick and strong band about its neck. This bundle of muscular fibres is called the *pyloric valve.* (Fig. II. 3.) When it contracts, it binds itself around the neck of the stomach, and closes the passage, as a string closes the mouth of a bag.

9. The duty of this valve is to close the aperture between the stomach and the alimentary canal, and retain the food and the gastric juice within the stomach during the process of digestion; and as fast as any part of the food is converted into chyme, the valve relaxes, the door opens, and the digested matter passes out.

10. But this door-keeper is endowed with a kind of sensibility, so that it opens only when the chyme or digested food is presented to it, and closes when the undigested food is offered, and refuses to let it pass through.

11. *Easy and Difficult Digestion.*— If the stomach is in good health, and if proper quantity and quality of food is taken, the work of

digestion of an ordinary meal is easily finished, and the body feels comfortable during, and refreshed after, the process. But if the meal is excessive, or of an indigestible nature, this process is laborious, painful, and protracted.

CHAPTER VII.

TIME AND CONDITIONS OF DIGESTION.

TIME.

1. THE time required for converting food into chyme, and for emptying the stomach of its contents, differs in different persons and circumstances. One kind of food is more easily dissolved than another; and some persons can digest any definite kind of food more readily than others; some are even oppressed by that which another digests with ease and comfort.

2. There is also a difference in regard to the digestive power in the same person in different states of health.

DR. BEAUMONT'S OBSERVATIONS.

3. A young soldier, Alexis St. Martin, had a hole through his side and into his stomach, in consequence of a gun-shot wound. Through

this aperture the whole process and progress of digestion could be observed. Dr. Beaumont availed himself of this opportunity to learn the time required to convert various kinds of food into chyme, and to empty the stomach of its contents.

4. The following table, which is taken from a larger table in Dr. Beaumont's work, shows the time required by St. Martin for the digestion of the most common articles of diet.

5. TABLE.

Articles.	Preparation.	Time Hr.	Time Min.	Articles.	Preparation.	Time Hr.	Time Min.
Apples, sour, hard,	Raw,	2	50	Eggs, fresh,	Fried,	3	30
———, sweet, do.	Raw,	1	30	Fowl, domestic,	Boiled,	4	
Beans, pod,	Boiled,	2	30	Lamb, fresh,	Broiled,	2	30
Beef, fresh, lean, rare,	Roasted,	3		Liver, beef's, fresh,	Broiled,	2	
——— steak,	Broiled,	3		Meat, hashed with vegetables,	Warm'd,	2	30
——— salted, old, hard,	Boiled,	4	15				
Beets,	Boiled,	3	45	Milk,	Raw,	2	15
Bread, wheat, fresh,	Baked,	3	30	Mutton, fresh,	Roasted,	3	15
———, corn,	Baked,	3	15	Pork, fat and lean,	Roasted,	5	15
Cabbage, with vinegar,	Raw,	2		——— steaks,	Broiled,	3	15
				Potato, Irish,	Boiled,	3	30
———,	Boiled,	4	30	———,	Roasted,	2	30
Cake, sponge,	Baked,	2	30	Rice,	Boiled,	1	
Cheese, old, strong,	Raw,	3	30	Sausage, fresh,	Broiled,	3	20
Chicken, full-grown,	Fricas'd,	2	45	Soup, beef, vegetables, and bread,	Boiled,	4	
Codfish, cured, dry,	Boiled,	2					
Corn, green, and beans,	Boiled,	3	45	Tripe, soused,	Boiled,	1	
				Turkey, domestic,	Roasted,	2	30
Custard,	Baked,	2	45	Turnip, flat,	Boiled,	3	30
Dumpling, apple,	Boiled,	3		Veal, fresh,	Fried,	4	30
Eggs, fresh,	Boiled,	3					

6. As the digestive power differs in different persons, this table must be received only as an approximation of the average periods required for the digestion of these articles,

Conditions of Digestion.

7. *Effect of Drink with Food.* — The gastric juice has just the right strength to dissolve food. If it is diluted with other liquids, it is weakened, and its power of digesting food is diminished. Nature seems to have provided for this; for, according to the observations of Dr. Beaumont, the first work of the stomach, after eating, is to relieve itself of the liquid that has been drunk with the meal.

8. Drink, therefore, taken with our meals, would appear to retard the digestive process by weakening the dissolving fluid, or, at least, to suspend it until the other fluid is carried away. Dr. Warren thinks, that very little drink is at any time necessary, and that the natural fluids of the mouth and the stomach — the saliva and the gastric juice — are sufficient to moisten and to dissolve the food.

9. *Effect of Mastication.* — The stomach is merely a soft and fleshy bag, and cannot do the proper work of the teeth upon the food. It is not fitted to crush hard substances or divide large morsels; and if it is required to digest such matters, it can only do so by a long and painful labor.

10. If large pieces of food are swallowed, the gastric juice can reach only their outer layers; and when these are softened, they are removed by the motions of the stomach, and then another layer is exposed to the fluid; and this process continues, until all the layers are successively dissolved and removed.

11. *Effect of Texture.* — If the food is of a hard, compact, or adhesive texture, such as dry or tough meats, or heavy bread, the gastric juice does not easily penetrate and dissolve it, and it must require a longer time to be digested than the tender and soft meats, and light and porous bread.

12. When any indigestible matters are eaten, the stomach endeavors to dissolve them; and after laboring in vain to convert them into chyme, it attempts to thrust them out. But the pyloric valve refuses to let them pass. Then, again, the stomach attempts to digest the insoluble food, and again tries to thrust it out. For a long time, the valve closes with a painful firmness, until at last it yields, and the crude matters pass through it. This is the cause of the distress which some persons feel two or three hours after eating indigestible food.

3*

CHAPTER VIII.

INTESTINAL CANAL.—CHYLE.

Coats of the Canal.

1. The alimentary or intestinal canal, like the stomach, is composed of three coats — the outer or *peritoneal* coat, the middle or *muscular* coat, and the inner lining or *mucous* membrane.

2. The *peritoneal coat* gives strength and support to this organ, and, being attached to the back-bone, holds the whole canal in its proper place.

3. The *muscular coat* has the power of contraction, and can be expanded. Its fibres expand to give room for the contents within the canal, and by contracting they draw it down close upon them, and leave no vacant room. After the food enters this canal, the circular fibres of the muscular coat contract successively, one after another, from the upper end at the stomach downward, and thus they press the contents onward.

4. The inner or *mucous coat* lines the whole canal. It is thick, soft, and spongy, and lies

loosely in folds or wrinkles. This coat prepares a slimy or mucous fluid, that covers its surface, and protects it from any irritation that might be caused by the matters within it.

Lacteal Absorbents and Ducts.

5. The mucous coat is filled with minute tubes, that open upon its surface, and run outward through the walls of the canal. They unite together and form larger tubes, which again unite, like the branches of a tree; and at last, all are joined in one trunk, which is called the *thoracic duct*. This duct passes along the back-bone and opens into the great vein near the heart, in the chest or *thorax*.

6. The numberless mouths on the inner surface of the alimentary canal, the little tubes that lead from them, and the great tube that connects them with the blood-vessels, constitute what is called the *lacteal system*. The mouths are called the *lacteal absorbents*, because they absorb or suck up the milky or lacteal chyle.

Chyle.

7. The chyme, at first, includes not only the nutritious portion of the food, which may be converted into blood, but the innutritious por-

tion, which is useless. Soon after entering the intestinal canal, these various elements are separated from each other. The nutritious part, called *chyle*, is white, mild, and bland, like milk. It is the same in all persons, however different may have been the food which they have eaten.

8. *Waste Portion of Food.*—The innutritious or waste portion is not only useless to, but a burden upon, the system; and if not frequently,—daily, or nearly as often,—carried from the body, it becomes a source of irritation and disturbance, not only to the digestive organs, but to the nervous system, and to the whole frame.

9. *Proportion of Chyle.*—The chyle alone nourishes the body. The proportion of this nutritious matter, which can be taken from the chyme, varies according to the character of the food, the method of its preparation, the completeness of the mastication, and the digestive power.

10. *Course of the Chyle.*—After the chyle is separated from the refuse matter, it is absorbed or sucked up by the lacteal absorbents, which cover the inner surface of the canal. Each one of these is almost inconceivably

small; yet they are so numerous, that the whole together absorb all the nutritive parts of the food we eat.

11. The chyle passes through the minute tubes into the larger branches, and thence, through the lacteal duct, into the blood-vessels near the heart.

FOUR PROCESSES OF DIGESTION.

12. These are the several established processes of digestion. 1st. Mastication and moistening with saliva in the mouth. 2d. Digestion and conversion into chyme in the stomach. 3d. Separation of the nutritious parts from the refuse matters in the intestinal canal. 4th. Removal of the chyle from the digestive organs to the blood-vessels.

13. These successive processes are all necessary for the final work of nutrition. If the mastication is imperfect and the food is swallowed without due preparation, the digestion will be imperfect, and less nutritive matter will be extracted from it.

C

CHAPTER IX.

CONDITIONS OF EATING.

1. *Supply of Food.*—THE digestion of food in the stomach is a natural process. The preparation of food before it enters the mouth is an artificial one, which is assigned to man. He is appointed to select, supply, and prepare the materials which will nourish him.

2. The supply of food must have reference to its quantity and quality, to the frequency and the hours of eating, to the constitution, temperament, health, habits, and age of the person, to the season of the year, and to many other circumstances.

APPETITE.

3. Healthy appetite or hunger, which shows that the body needs more nourishment and the stomach is ready to digest more food, informs us how often and how much we should eat. But this guide can be followed only when we eat slowly and swallow food no faster than the gastric juice is prepared, and suspend our eating when this fluid ceases to flow.

4. *False Appetite.* — The natural appetite becomes diseased by improper stimulation. The mere craving for exciting food — for that which is pleasant to the palate — may exist without a corresponding want of nourishment in the frame, or digestive power in the stomach. This is a false and unnatural appetite, and, if it is indulged, it tempts us to overload the stomach, or to eat indigestible food.

5. *No Digestive Power without Appetite.* — There may be an apparent appetite without digestive power; but, on the contrary, whenever the body wants more nourishment, and the stomach is ready to digest it, they show, through the natural appetite or hunger, the propriety of taking food.

6. The absence of hunger then indicates the absence of digestive power; and consequently, if food is eaten when it is not needed, it is not easily converted into chyle in the digestive organs, and does not nourish the body in the best manner afterwards.

7. *True Appetite a Guide in Eating.* — By careful attention to the natural cravings, we are generally informed by our appetites as to the quantity of food which we may take, and the times of eating, and the intervals of our meals.

Quantity of Food

8. The quantity of food necessary varies with the age, health, and habits of the person, and the season of the year. As all action increases the waste and changes of atoms, the laborer needs more nourishment, and has a better appetite, and can eat and digest more food than the man who exercises but little.

9. Children and youth who are growing in stature, convalescents who are regaining their lost flesh after sickness, and healthy persons, who are growing fat at any time must eat more food than others, to supply the want of nutriment that is created by the increasing size or stature of the body.

Rest after Eating.

10. Healthy and vigorous digestion requires all the energies of the body in its earliest stages. If these energies are directed to any other operation,—to the labor of the hands or the feet, or to any great exertion of the brain,— the stomach acts with less power and effect, and the food is not so well digested. Rest, therefore, for a short period after eating, favors the conversion of the food into chyle.

REST BEFORE EATING.

11. Whenever the body is fatigued or exhausted with any labor, all the parts and organs partake, in some degree, of the languor, and all — the stomach as well as the limbs — are then somewhat unfitted for exertion, and unable to work with the most successful energy. Digestion then cannot be well performed, when the body is fatigued, and food should not be taken until the body has enjoyed some rest after hard labor.

12. This relaxation of labor before eating, and rest afterwards, are necessary for the perfect digestion of food, and for the invigoration of the body. And it is especially necessary for those laboring men who wish to accomplish the most with their exertions, and need for this purpose to gain the greatest power from their nourishment. This cannot be done in the short space — the few minutes or the half hour — frequently allowed for the laborer's meals.

CHAPTER X.

QUALITY AND FREQUENCY OF FOOD.

Quality of Food.

1. The various kinds of food affect the living body variously. The two natural divisions of food — the animal and the vegetable — differ most in their characters and in their effects upon the human constitution.

2. *Meat* contains the most nutriment, and is the most stimulating. It is therefore eaten more by active laborers, and by persons exposed to cold, than by the sedentary, and those who work in warm shops or live in warm rooms. It is eaten more in winter than in summer, and more by the inhabitants of the frigid zone than by those who dwell within the tropics.

3. The stimulating diet of the people of the frozen regions would produce fever and other derangements if it were used by the inhabitants of the burning regions near the equator. On the contrary, the cooling diet of the tropical climates, would be insufficient to maintain the heat and support the strength of the animal body near the poles.

4. The *differences of constitution* require cor-

responding differences of food. The nervous and the excitable are better supported on a mild and cooling diet, and the calm and the dull need to be nourished and strengthened by a stimulating diet. The former do better with a large proportion of bread and vegetables, and the latter with a large proportion of meat.

Condiments and Stimulants.

5. The natural powers of the stomach, and strength of the gastric juice, are sufficient, in a state of health, to digest all suitable kinds of food; and these need no artificial aid from stimulation. Condiments, spices, wines, alcoholic liquors, and fermented drinks excite the stomach to an unnatural activity, and often quicken the digestive process; but this over-action exhausts the powers of the digestive organs, and ultimately brings on weakness, and sometimes disease.

6. *Natural Appetite.* — The natural taste is keen, and discriminates and enjoys the differences and the qualities of simple food. But if the palate is excited by stimulating condiments, spices, or drinks, it loses its acuteness of perception, and its capacity for enjoying simple things; and then it requires a more stimulating diet to excite it.

7. *Food should be agreeable.* — Having selected food of simple and digestible nature, and of such quality as will meet the wants of the body, it is also proper that it should be made pleasant to the natural appetite. Within these limits, pleasant and savory food is not only allowable, but very proper, for the healthy stomach will best digest the food which the unperverted appetite enjoys.

8. *No Rule of Diet for all.* — Men differ so much in their constitutions, their health, and their habits, that it is impossible to prescribe any single rule of diet that all may safely follow, or that even the same person may obey in the various conditions of his body or the various periods of life. Every one must study the law of his being, and examine his own constitution, temperament, habits, and exposures, and then adapt his food — both its quantity and quality — to his own peculiar wants.

FREQUENCY OF MEALS.

9. About six hours after a proper meal, the nutritive particles of the blood are so much reduced, that nutrition begins to languish, and then the body feels the want of another supply of nutriment. In the same period, the stomach

digests the last meal, and has time to rest and regain strength for another work. Then the concurrence of these two conditions indicates itself by hunger.

10. *Times of Eating.* — The usual meals — three times a day — are in accordance with the natural wants and powers of the human body. As the night is spent mostly in inaction, the interval between the evening and the morning meal must be greater than that between the meals of the active part of the day. Yet the body wants some refreshment before it engages in active labor in the morning.

11. The *breakfast*, therefore, should be taken soon after rising. The *dinner* should follow the breakfast, and the *supper* follow the dinner, in about six hours.

12. But if the supper is taken very near the time of sleeping, the labor of the digestive organs may interfere with that composure of the body which is necessary for perfect rest, and thereby interrupt sleep. The supper, then, should be taken at least two or three hours before bed-time, so that the food may be nearly digested when we lie down; and then the stomach will be ready to rest with the other organs of the body, and all sleep together.

CIRCULATION OF THE BLOOD.

CHAPTER XI.

APPARATUS OF CIRCULATION.

1. The sole object of eating food is to furnish chyle for the blood; and the whole purpose of the blood is to supply the body with the materials necessary for its nourishment.

2. The chyle is mixed with the blood in the great vein, and then they flow together into the heart. After due preparation, this new mixture of the chyle and the old blood is sent from the heart to all parts of the body, by the apparatus for the circulation of the blood.

3. The apparatus of circulation consists of the heart, the arteries, the capillaries, and the veins.

4. The *heart* is the central organ or engine that propels the blood through the arteries.

5. The *arteries* proceed from the heart, and run all over the body, even to its farthest and

minutest parts, and carry the blood in their channels.

6. The *capillaries* receive the blood from the arteries, and transmit it to the veins.

7. The *veins* receive the blood from the capillaries, and carry it back to the heart.

Heart.

8. The heart is placed in the chest, behind the lower part of the breast-bone. It is a hollow, fleshy bag, composed of muscular fibres, like the middle coat of the stomach, or middle layer of tripe. These fibres contract like the leech; and, when they contract, they lessen the internal cavity of the heart, and press out whatever may be contained in it.

9. *Divisions.* — The heart is divided into two parts, the right and the left, which have no direct communication with each other, for they are separated by an impervious wall. Each of these divisions is subdivided into two smaller apartments — the upper called the *auricle*, (Fig. VI. 1, 6, p. 50,) and the lower called the *ventricle*, (Fig. VI. 2, 7.) There is an open passage-way from the upper to the lower chamber, through which the blood passes from the auricle above to the ventricle below.

44 CIRCULATION OF THE BLOOD.

10. *Valves.* — In the passage-ways between the auricle above and the ventricle below, on

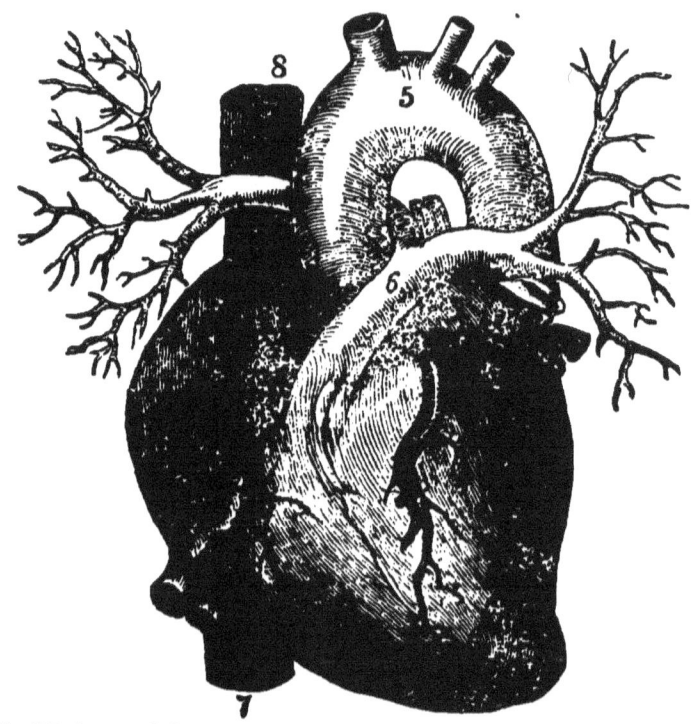

FIG. IV. — *Heart.*

1. Right auricle.
2. Left auricle.
3. Right ventricle.
4. Left ventricle.
5. Great artery carrying the blood from the left ventricle to the body.
6. Artery carrying the blood from the right ventricle to the lungs.
7, 8. Great veins carrying the blood to the heart from the body.

both sides, there are valves placed, answering the same purpose as the valve of a pump-box, which opens to let the water pass upward, but closes and prevents its passage downward. In like manner, these valves open and allow the blood to pass down from the auricles to the ventricles, but close when these lower cavities are filled, and prevent the return of the blood to the upper cavities.

11. There are also valves between the ventricles and the arteries, that lead out from them. These allow the blood to pass out of, but not to return into, the heart.

CHAPTER XII.

BLOOD-VESSELS AND THEIR ACTION.

1. *Arteries and Veins.* — THERE are four sets of vessels or tubes connected with the heart. Two of these sets, called the *arteries*, carry the blood out from the ventricles; one carries it from the right side to the lungs, (Fig. VI. 8, 8, p. 50,) and the other carries it from the left side to the whole body, (Fig. VI. 3, 3.) Two other sets, called the *veins*, bring the blood back to the auricles; one brings it from the whole body to the right side, (Fig. VI. 5, 5,) and the other

brings it from the lungs to the left side of the heart, (Fig. VI. 10, 10.)

2. Each of these sets of blood-vessels has a single and large tube at and near the heart; and, as these tubes extend from it, they are divided into many and smaller branches; and these again are subdivided and multiplied,—increasing in number and diminishing in size,—until the little arteries and veins are countless and almost invisible. In this division and multiplication, they resemble a tree which has a single and large trunk at one end, and many small branches at the other.

3. Following the course of the moving blood, the arteries are said to lead out from the heart to every part of the body, beginning with the large trunk and ending in the innumerable minute branches. The veins are said to lead from all the parts of the body to the heart, beginning with the minute branches, and ending in the great trunk at the heart.

4. Fig. V. shows the distribution of the arteries of the face. The blood-vessels are distributed through the other parts of the body in the same manner. The distribution of the veins through the body is similar to that of the arteries. The trunks of these two sets of vessels do not go together, side by side, but their

minute branches and terminations reach the same points.

Fig. V.—*Arteries of the Face.*

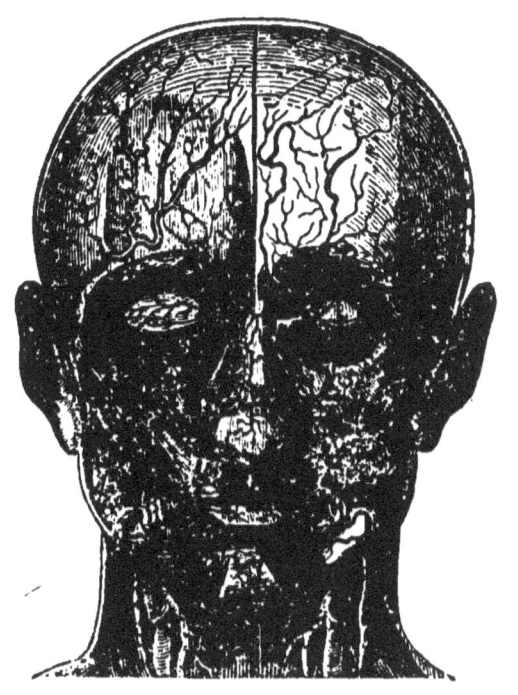

5. These two sets almost meet, with their trunks at the heart, and with their minute terminations in every part of the body. The large trunk of the veins opens into the upper chamber of the right side, and the large trunk of the arteries opens out of the lower chamber of the left side, and there is an impervious wall between them. Between the minute terminations of the arteries and the minute beginnings

of the veins there is a third set of blood-vessels, still more minute, called the *capillaries*.

Course of the Blood in the Body.

6. The heart is the propelling engine in the circulatory system. By its contractions it presses the blood out of its cavities, and forces it into the arteries. This blood flows from the left ventricle into the arteries, and through the arteries into the capillaries, and through the capillaries into the veins, and lastly, through the veins back to the right auricle. This is the *general circulation*, which sends the blood through the whole body. This blood goes out from the left side, and returns to the right side of the heart.

Course of the Blood in the Lungs.

7. Before the blood can pass from the right side to the left side of the heart, it must pass through the lungs, by what is called the *pulmonary circulation*. The arteries that lead from the right ventricle are divided and distributed throughout the lungs, in the same manner as the arteries of the general circulation are distributed throughout the body. The minute and numberless extremities of the pulmonary

arteries in the lungs meet with, and open into, veins of similar number and size; these countless little veins are united and gathered into larger and larger trunks, and finally terminate in the left auricle, or left upper chamber of the heart. (Fig VI. 7, 8, 9, 10, p. 50.)

Double Circulation.

8. The scarlet blood passes from the left auricle (Fig. VI. 1) to the left ventricle, (Fig. VI. 2,) and from the left ventricle, through the arteries, (Fig. VI. 3, 3,) to the capillaries in all the upper and the lower parts of the body, (Fig. VI. 4, 4,) where its color is changed to purple, and returns, through the veins, (Fig. VI. 5, 5,) to the right auricle, (Fig. VI. 6,) and from the right auricle to the right ventricle, (Fig. VI. 7.) Again, this dark or purple blood passes out from the right ventricle, through the pulmonary arteries, (Fig. VI. 8, 8,) into the lungs, where its color is changed from purple to scarlet; and then it passes through the minute vessels (Fig. VI. 9, 9) into the pulmonary veins, (Fig. VI. 10, 10,) and through these veins back to the left auricle, (Fig. VI, 1.); and thus the double circulation — the general and the pulmonary — is completed.

CIRCULATION OF THE BLOOD.

Fig. VI. — *Double Circulation.*

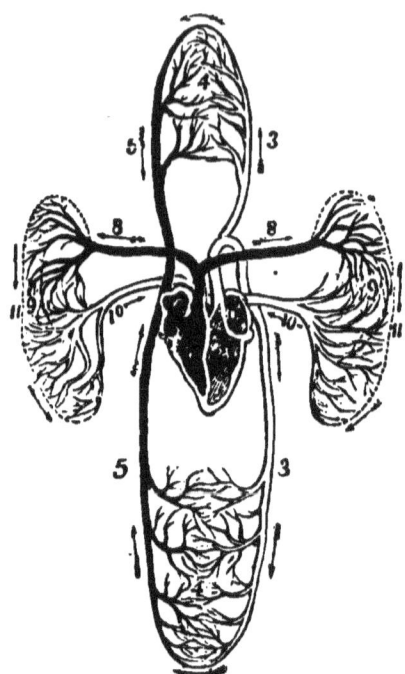

1. Left auricle.
2. Left ventricle.
3, 3. Arteries leading to the upper and lower parts of the body.
4, 4. Capillaries.
5, 5. Veins leading from the upper and lower parts of the body.

6. Right auricle.
7. Right ventricle.
8, 8. Pulmonary arteries.
9, 9. Minute blood-vessels in the lungs.
10, 10. Pulmonary veins.
11, 11. Lungs.
The arrows show the course of the moving blood.

ACTION OF THE HEART.

9. In a state of health, the heart beats or contracts, in a grown person, about seventy-five times a minute, and sends, at each contraction

or pulsation, about half a gill of blood from its left side to the body, and as much from its right side to the lungs.

10. When the heart sends this blood into the arteries, they expand and beat. This beating of the arteries is called the *pulsation*, and is felt at the wrist, and at the neck, and wherever else they come near the surface.

11. The action of the heart is affected by various circumstances; it is increased in most diseases, and diminished in some others. It is increased by exercise, and especially by rapid running, by stimulation of spirits, and by excitements of the mind. It is somewhat diminished on lying down and in sleep.

Quantity of Blood flowing.

12. The quantity of blood in the body of a man of average size is about twenty-eight pounds. Two ounces of this blood pass through the heart and through the body seventy-five times a minute, and the whole four hundred and forty-eight ounces pass through once in three minutes. More than sixteen hundred gallons of blood are thus received into, and sent out of, the heart in the course of a day.

NUTRITION.

CHAPTER XIII.

GROWTH.

1. The blood flows through the body to carry the materials which are to nourish its parts and supply its waste. In early years, when the body is growing in stature, and at other periods of life, when it is gaining in weight, the increase is made by the addition of new particles from the blood.

Changes of Particles.

2. During the whole of life, the particles of the body are constantly changing. When an atom of matter is taken from the blood and converted into flesh, it receives the principle of life and the peculiar properties of the part or organ in which it is deposited. With this living power and these properties, it acts in connection and concert with the rest of the

part or organ to which it is joined; but after a period, it is exhausted and loses its power; then it is dead and can act no more.

3. When an atom of flesh is dead, it is absorbed or taken away by a little vessel called an *absorbent,* and its place is supplied by a fresh and living atom from the blood. This new atom, like the one which went before it, fills its place, and acts there for a time; and then it dies, and gives its place to another.

4. This change from life to death, and of dead atoms for living atoms, is constantly going on in all the parts of the body; and thus all our flesh is continually dying; and it is also continually renewed with living particles.

5. Here, then, is a double necessity for nutrition and for food — 1st, the growth and increase of flesh in early years, and occasionally at other periods of life; and 2d, the changes of atoms at all periods from birth to death.

6. All the parts of the body, however different they may be, — the flesh and the bone, the eye and the hair, — are formed out of one blood. The work of nutrition is performed in or by the capillaries, while the blood is passing through them from the arteries to the veins. These little vessels, in each part, select from

the blood the very elements that are needed in that part, and in the proportion that is necessary to form the kind of flesh or living substance that is to be renewed. In the bone they select the elements that form bone, and in the muscle they select those which form muscle, &c.

COÖPERATION OF NUTRITION AND ABSORPTION.

7. The *capillaries*, or nourishing vessels, and the *absorbents* are distributed to every part of the living frame. They work together in concert, and keep the body at about its usual size. In grown persons, one set of vessels deposits as many atoms as the other takes away; otherwise the body might grow very large, or waste entirely away.

8. In childhood and youth, when the body is growing, and at other times, when it is gaining flesh, nutrition predominates, and then more atoms are deposited than are taken away. In old age frequently, and at other periods, when the body is losing flesh, absorption predominates, and then more atoms are taken away than are deposited. But during the middle periods of life, these operations usually balance each other, as many atoms being added

as are absorbed, and thus the body neither gains nor loses in weight.

9. *Effect of Exercise.* — Exercise of any part exhausts the life of its atoms, and causes more rapid absorption of the old, and a necessity for more rapid nutrition to supply new atoms. The most active have the greatest waste, and consume the most blood, and need the most nutriment to supply this loss. Laboring men, who waste much flesh and consume much blood, must eat more food to nourish their frames than sedentary or idle persons.

10. *Effect of Age.* — Both nutrition and absorption are more rapid in early years, and slower in old age, than in the middle periods of life. Both are more rapid in the laborious than in the inactive. The atoms remain in the young and the active a shorter time than in the old and sedentary. The atoms that compose the flesh of the former, who exercise much, are fresher and newer than those which compose the flesh of the student, who exercises little

EFFECT OF NUTRITION ON BLOOD.

11. In this process of nutrition, which is performed in or by the capillaries, the blood loses some of its nutritious particles, and is therefore

poorer in the veins after nutrition than it is in the arteries before it. Its color is also changed, by the same process, from the bright scarlet, as seen in the arteries, to the dark purple, as seen in the veins. (Fig. VI. 3, 3, 5, 5, p. 50.)

Dead Atoms of Flesh.

12. When the exhausted and dead atoms, which constitute the waste matter of animal life, are taken from their places in the body, they are carried into the veins, and mixed with the venous blood. Then both of these are carried together through the small veins to the larger branches, and through these to the great trunk which opens into the right side of the heart, and thence they are sent to the lungs.

13. The dead particles, that are thus taken into the veins and carried to the heart, amount to several ounces a day. If they should remain in the body, they would soon fill and overload the blood-vessels and cause disturbance and finally death. But provision is made for their removal through the lungs by means of respiration.

RESPIRATION.

CHAPTER XIV.

CHEST.

Venous Blood.

1. The purple blood, when it enters the right auricle from the veins, consists of three parts: 1st, the residue of the arterial blood which has not been consumed in the supply of nourishment to the textures throughout the body; 2d, the waste or the dead atoms of flesh that have been absorbed from the textures and carried into the veins; 3d, the new chyle which has been received from the digestive organs.

2. This blood, in the right side of the heart, cannot serve the purposes of nutrition; if it is carried through the arteries to the textures, it will not only fail to nourish but injure the organs of the body. The old blood is so much reduced by the loss of its particles which have

been consumed in nutrition, that it cannot again supply the new atoms to make new flesh; the new chyle is not yet completed and ready to be converted into flesh; and the dead atoms will cause irritation and disease, if they are carried again through the arteries.

3. Before this blood can nourish the body, it must be relieved of its dead atoms, that would be injurious, and the old blood must be renewed and strengthened by the perfect union of the chyle with it. Both these things are done in the lungs, by means of respiration and the air.

Lungs.

4. The lungs are placed in the chest, in the upper part of the body. They are furnished with tubes, to receive the air from abroad; and with vessels, to receive the blood from, and to carry it back to, the heart. The lungs are enclosed in a movable case, which can be expanded for the admission of air into the air-tubes, and contracted to expel it from them.

Chest.

5. The chest constitutes the upper part of the trunk of the body. It extends from the neck to the abdomen. It is made of a bony

framework on its upper part and on its sides, and of a fleshy layer of muscle below. There are also layers of muscular flesh between the bones at the sides.

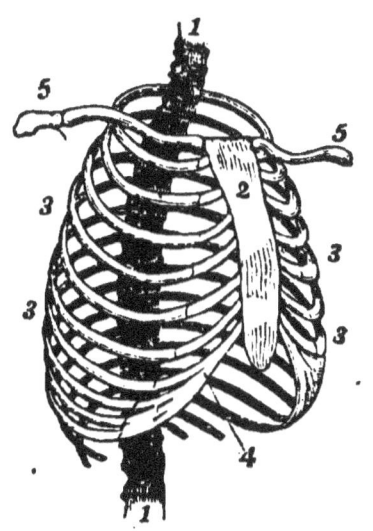

FIG. VII. — *Chest.*

1. 1. Back-bone.
2. Breast-bone.
3, 3, 3, 3. Ribs.
4. Cartilages.
5, 5. Collar-bones.

6. The framework of the chest includes that part of the back-bone, or *spine*, which extends from the neck to the loins, (Fig. VII. 1, 1;) the breast-bone, or *sternum*, which lies front, and extends from the neck to the pit of the stomach, (Fig. VII. 2,) and the twenty-four *ribs*, which cover the sides and most of the front and the back of the chest, connecting the back-bone with the breast-bone. (Fig. VII. 3, 3, 3, 3.)

7. The back-bone is the pillar on which the chest is supported. There are twelve ribs on each side. They are curved and bent, in a somewhat circular form, from the spine to the sternum, and give the rounded form to the chest. Their posterior ends are attached to the back-bone, by a joint that allows the rest of the rib to move. They are attached to the sternum, or breast-bone, in front, by means of a stiff substance or gristle, called *cartilage*. (Fig. VII. 4.)

8. The posterior end of the rib is fixed, and cannot move from its place. But it can roll in its joint or socket in the back-bone, as one part of a gate hinge moves in the other. The anterior end of the rib is attached more loosely to the breast-bone, which is not fixed and immovable, like the back-bone. The whole anterior portion of the ribs can be moved upward and downward on their posterior ends or joints as the lid of a chest can be raised on its hinges. When the ribs are raised or depressed, the sternum and the whole anterior framework of the chest are lifted up or fall together.

9. The course of the ribs from the back-bone to the breast-bone is not horizontal, but they run obliquely around the chest, inclining downward, so that their anterior ends at the

breast-bone are lower than their posterior ends at the spine.

10. The oblique direction of the ribs gives less diameter to the chest than a horizontal course would; and when the anterior ends are raised to a level with the posterior ends, they must carry the breast-bone farther forward, and to a greater distance from the back-bone, and thus leave a wider space between them and a larger cavity within.

11. The natural position of the ribs, when at rest, is oblique; and then the sternum stands nearest to the back-bone, and the cavity of the chest is reduced to its smallest size. When the ribs move, their shafts rise toward the level of their posterior and fixed ends, and carry their anterior ends and the sternum forward and outward, and thus enlarge the diameter of the chest and the size of its cavity.

Fig. VIII.

12. The ribs surround the chest as a large hoop surrounds a small cask lying obliquely. (Fig. VIII.) If the lower side of the large hoop be raised from the point 3 to 2, it will allow the staves to spread and enlarge the cavity of the cask.

CHAPTER XV.

MUSCLES OF THE CHEST.—DIAPHRAGM.

Muscles of the Chest.

1. The movements of the ribs are caused by a set of muscles provided for the purpose. These muscles are composed of lean fibres, like the heart or the middle coat of the stomach (Chap. IV. § 5,) and have a similar power of contraction.

2. Some of these muscles lie between the ribs, running from one to the other. These are called *intercostal muscles*. Other muscles, called *spinal muscles*, are fastened by their upper ends to some part of the spine, and by their lower ends to some of the ribs.

3. *Expansion of the Chest.*— When these muscles contract, they draw their ends toward each other, and of course draw up the ribs, which are movable, toward the upper bones, which are fixed.

4. The intercostal muscles, lying between the ribs, draw the lower ribs, which are loose, toward the upper ribs, which are held in their places by the muscles which come from the

neck and spine. By the combined actions of these muscles, — the spinal and intercostal, — the ribs are raised, the chest is expanded at its sides, and its cavity is enlarged.

5. *Contraction of the Chest.* — These muscles have only a momentary action. They contract and raise the ribs, and then relax immediately, and allow the ribs to fall and return to their natural oblique position; and thus the chest is again contracted. This is done partly by the elasticity of the cartilages, and partly by the action of the abdominal muscles which aid in drawing the ribs downward.

Diaphragm.

6. The ribs bound the sides on the upper part of the chest, and surround it with a bony encasement. The bottom of this cavity is bounded by a broad and thin muscle, or layer of lean flesh, called the *diaphragm.* This muscle extends across the bottom of the chest from side to side and from front to back. Its edges are attached to the lower side of the lowest ribs, to the lower extremity of the sternum, and to the back-bone.

7. The diaphragm is not flat, nor does it lie in a horizontal position. Its anterior edge, which

is attached to the sternum, is much higher than its posterior edge, which is attached to the spine. Its centre is raised above its edges, and arched upward into the cavity of the chest, like a dome, or like the bottom of a glass bottle which is turned inward.

8. When this muscle is at rest, it is expanded; and then the point of its arch rises within the chest as high as the fourth rib from the top; but when it is in action and contracted, the point of this arch is drawn down as low as the seventh rib from the top.

9. At the same time that the intercostal and spinal muscles raise the ribs and the sternum, and enlarge the cavity of the chest above, the diaphragm contracts, draws down its arch, and enlarges the chest below. The combined actions of these muscles above and of the diaphragm below perform the mechanical part of inspiration. They enlarge the cavity of the chest, make room for the lungs to expand and for the air to pass in, and thus we get our breath.

10. The diaphragm, like the muscles of the ribs, has only a momentary action. It soon ceases its contraction, and then relaxes, and allows its arch to be thrown upward into the chest.

11. The diaphragm (Fig. III. 3, 3, p. 23,) is

immediately above the stomach and the other digestive organs, (Fig. III. 6, 7, 8,) which fill the whole cavity of the abdomen. The *abdominal muscles* form the outer, or front and lower walls of the abdomen, and cover all its contents. They extend from the lower edge of the chest above, to the pelvis, or the great bone that runs across the hips, below. (Fig. XIV. 4, 4, p. 115.)

12. When the diaphragm contracts and draws its arch down, it presses upon the stomach, the liver, and the other contents of the abdomen, and forces them downward; then the abdominal muscles yield to give room for them; and thus the abdomen is enlarged and extended forward and outward at every inspiration.

13. When the diaphragm ceases to act, and relaxes itself, the abdominal muscles contract, and, drawing their ends toward each other, they pull the ribs, which are movable, down toward the pelvis, which is immovable; by the same action, they press upon the digestive organs, and force them upward against the diaphragm, which ascends again into the chest and lessens its cavity.

14. The pressure of the abdominal muscles upon the diaphragm lessens its cavity below.

and the pulling of the ribs downward lessens it above. These combined actions, — the falling of the ribs and the elevation of the diaphragm, — constitute the mechanical part of expiration, and force the air out of the lungs.

CHAPTER XVI.

LUNGS.

1. The great purpose of all this framework of the chest, and its moving power, is to give place for, and motion to, the *lungs*. These are two large and spongy bodies, placed in the cavity of the chest, (Fig. IX. 1, 2,) one on each side, with the heart between them. (Fig. IX. 3.) They are soft, and, when filled with air, lighter than water. In the lower animals, they are called *lights*.

2. The object of the lungs is to bring the impure blood and the air from abroad together, so that the latter may purify the former by carrying off its load of waste atoms, and fit it to nourish the body. For this purpose, they are furnished

and almost filled with blood-vessels and air tubes, which compose most of their substance.

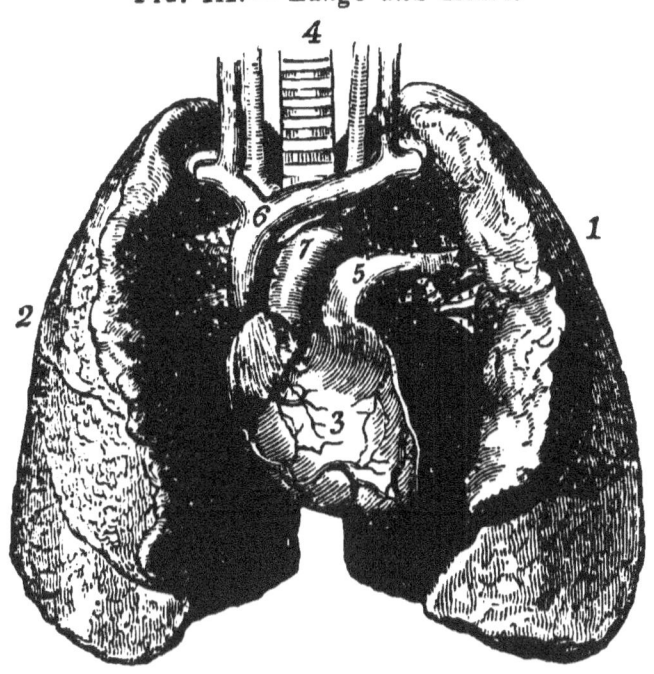

FIG. IX.— *Lungs and Heart.*

1. Left lung.
2. Right lung.
3. Heart.
4. Windpipe.
5, 6, 7. Great vessels going out of the heart.

WINDPIPE.

3. The system of air-vessels commences, at the back part of the mouth, with one tube, commonly called the *windpipe*, (Fig. X. 1, 2,) which passes through the neck to the chest. In the upper part of the chest, this air-tube is divided into smaller tubes, (Fig. X. 3, 3,) which

pass, one to the right and the other to the left lung. These tubes are again divided and sub-

Fig. X. — *Air-Vessels of the Lungs.*

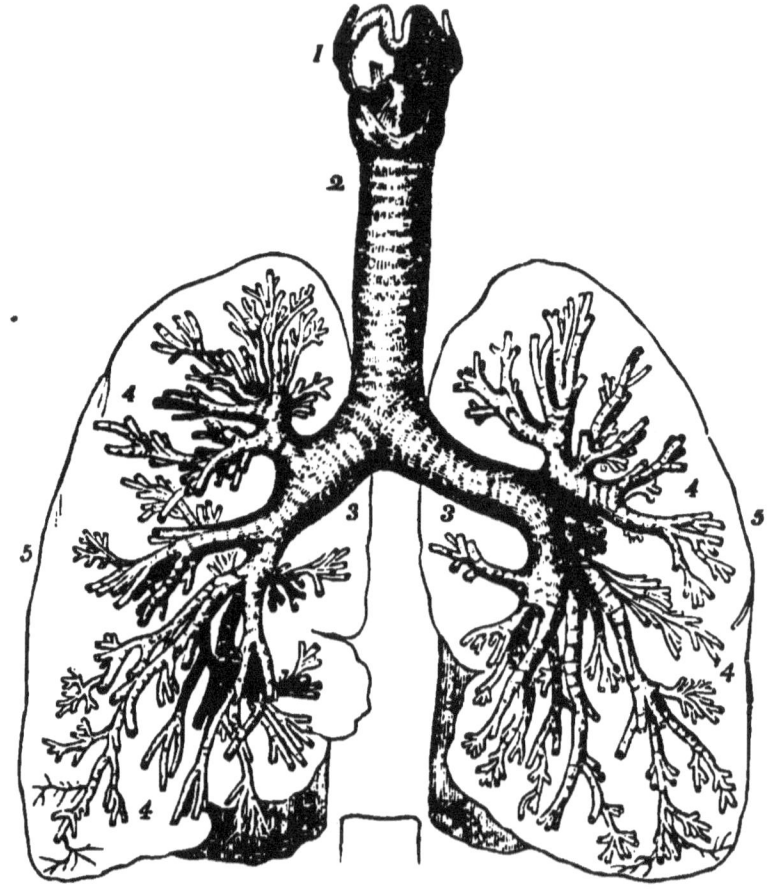

1. Larynx.
2. Trachea.
3, 3. Divisions to right and left.
4, 4, 4. Minute terminations.
5, 5. Outline of the lungs.

divided into still smaller and more numerous branches, until finally the minute branches terminate in very minute cells, (Fig. X. 4, 4, 4,) which are distributed through all parts of the lungs.

4. The windpipe, extending from the mouth to the chest, lies in the front part of the throat. It includes the *larynx*, (Fig, X. 1,) at its upper part, and the *trachea*, (Fig. X. 2,) below. The larynx opens into the back chamber of the mouth, through a narrow chink, called the *glottis*. The *epiglottis* is a little valve placed over this opening to prevent the entrance of any improper matters into the air passages.

Vocal Chords.

5. Just below the glottis, and within the mouth of the larynx, are placed some little bands of flesh, called the *vocal chords*. These chords extend across the sides of the channel or tube, and project somewhat into it. By means of muscular fibres, they may be drawn tightly or relaxed more loosely. They vibrate when the air passes over them, and thus produce the sound of the voice, which differs according to the tension to which they are drawn.

6. When the vocal chords are disordered by inflammation or otherwise, they produce a hoarseness of voice, as in common colds: or they may be so affected by ulceration or other

disease, as to entirely prevent the loud voice; and then the sufferer can only talk with the lips in whispers.

7. *Structure of the Air-Tubes.* — The larynx is composed of very firm cartilages, and includes that stiff and projecting part of the windpipe commonly called *Adam's apple*, in the upper part of the throat. The trachea is composed of stiff rings of cartilage in front, and of flesh behind. The larger air-tubes, within the lungs and chest, are also cartilaginous, and usually are open even when empty. The smaller air-tubes and the air-cells are soft and more loose, and are closed when not filled with air.

CHAPTER XVII.

PULMONARY AIR AND BLOOD-VESSELS

Pulmonary Air-Vessels.

1. *Mucous Membrane.* — The whole of the system of air-vessels, — the larger and the smaller tubes and the cells, — in the mouth and nostrils, and in the throat and lungs, is lined

with a *mucous membrane* similar to that which lines the stomach and alimentary canal.

2. In good health, this lining membrane is always moistened with *mucus*, its natural secretion. When it is inflamed by a common cold, this secretion is increased, and then it is thrown off by coughing or otherwise. A cold in the head increases it in the nasal passages; and a cold in the throat or in the lungs increases it in those parts.

3. *Coughing.* — The mucous membrane is made to bear the contact of air, but nothing else; any other gas or matter which is not adapted to its nature irritates it. Whenever any particle of food or drink gets within the glottis, an irritation is produced in the windpipe, and then nature uses the means which she has provided for the purpose, to expel the offending cause.

4. In these cases, the chest expands and the lungs inhale the air; then the muscles of expiration suddenly contract, and force the air out through the tubes, and drive the disturbing matter away. This is *coughing*.

5. Likewise, when any one breathes dust, smoke, or pungent gases, or when the mucous membrane prepares and throws into the air-ves-

sels an unusual quantity of mucus, or when the diseased matter of consumption gets into these passages, the same expulsive effort of coughing is made to force the irritating cause away.

6. When this membrane is irritated by disease or any other cause, coughing is excited to remove the irritation, although there is no matter to be removed. Sometimes this membrane prepares no mucus, and is dry; then the contact of the air irritates it, and causes a hard, dry cough.

PULMONARY BLOOD-VESSELS.

7. The lungs receive two sets of blood-vessels from the heart. One set, called the *pulmonary arteries*, (Fig. VI. 8, 8, p. 50,) passes from the right lower chamber, or right ventricle, to both lungs, and is divided into numberless branches, which are distributed throughout these organs. These carry the blood out from the heart.

8. The other set, called the *pulmonary veins*, (Fig. VI. 10, 10, p. 50,) is constructed and arranged in a similar manner, and connects all parts of the lungs with the left auricle or upper chamber of the heart. These carry the blood from the lungs back to the heart.

9. *Pulmonary Circulation.* — The blood is gathered from the whole body into the right

side of the heart. It is then forced out from the right ventricle, at every pulsation, through the pulmonary arteries into both lungs. The

Fig. XI.—*Interlacing of Pulmonary Blood and Air-Vessels.*

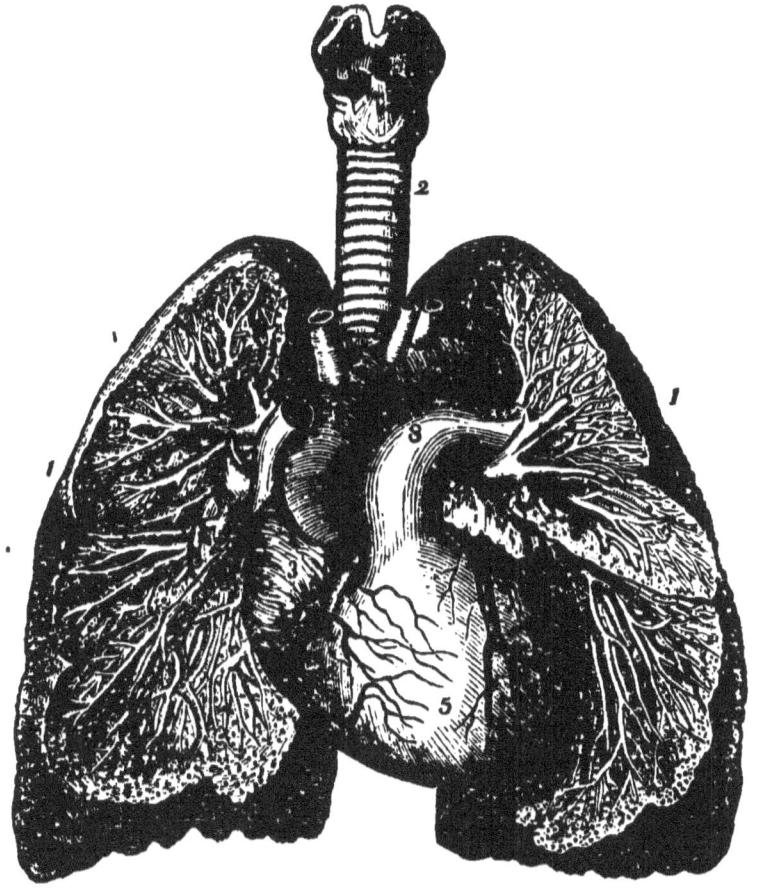

1, 1. Lungs.
2. Windpipe.
3. Right auricle.
4. Left auricle.
5. Right ventricle.
6. Left ventricle.

7. Aorta, or great artery leading to the body.
8. Pulmonary artery.
9, 9. Divisions of vessels in the lungs.

blood passes through the larger into the smaller trunks, and finally into the minute branches of the arteries; there it comes in contact with the air-cells, and is brought under the influence of the air contained in them. It then passes from the minute arteries to the minute veins, and through these into the larger venous branches and trunks, and finally into the left auricle, (Fig. VI. 8, 9, 10, p. 50.) When the blood enters the lungs, it is dark purple; but when it comes out of the lungs, it is bright scarlet. (Chap. XII. §§ 8, 9.)

10. At every pulsation, as much blood flows from the right side of the heart to the lungs as from the left side to the body; consequently, in a man of average size, in good health, about nine and a half pints flow from the right ventricle through the lungs, and back to the left auricle, every minute. (Chap. XII. § 12, 251)

CHAPTER XVIII.

RESPIRATION.

1. *Inspiration.* — The alternate operations of the muscles of the chest and of the diaphragm,

and of the muscles of the abdomen, produce respiration. The muscles of inspiration spread the ribs, draw down the arch of the diaphragm, and thus expand the chest and enlarge its cavity. The only inlet to this cavity is through the windpipe into the air-vessels of the lungs. When this cavity is enlarged, the air rushes in to fill the vacuum that otherwise would be made, and fills all the cells.

2. *Expiration.*—This inspiratory action lasts but a moment; for almost as soon as the ribs are raised and the arch of the diaphragm is drawn down, and the air let into the lungs, the muscles suspend their action, and allow the ribs to fall and the diaphragm to rise; at the same time, the abdominal muscles draw the ribs down, press the diaphragm up, and force the air out of the lungs.

3. In healthy adults, of average size, the successive and alternate actions of these two sets of muscles, producing inspiration and expiration, take place about eighteen times a minute, and expand the chest sufficiently to receive about forty cubic inches of air. But if the chest is small, and cannot expand to receive the due quantity of air, we breathe more rapidly to make up the deficiency.

4. *Meeting of the Blood and the Air.* —While the chest is expanding and contracting, and the air flowing into, and out of, the lungs, the heart is sending about half a gill of blood into them seventy-five times a minute, and as much returns from them in the same time.

5. The blood and the air are thus brought together in the lungs; the blood is in the minute capillaries, and the air in the little cells, with only an exceedingly thin film of flesh between them. This membrane is so constructed as to prevent the passage of liquids through it, but vapor and gases can pass through.

6. The blood is securely retained in its appropriate vessels, but the carbon, or the carbonic acid, and the hydrogen can go from them, through the covering, to the air-cells, and the air, or the oxygen in the air, can pass into the blood. The blood is thus brought within reach of the air, which relieves it of its waste or dead atoms, and perfects the new chyle, and fits it for the nourishment of the body.

COMPOSITION OF THE DEAD ATOMS.

7. *Carbon.* — The waste matter consists mostly of carbon and hydrogen. *Carbon* is a common element in nature. It is pure in the

diamond; it is very abundant in charcoal. It enters very largely into the composition of wood and all other vegetable substances. It is present in the various kinds of flesh, and in most animal matters. When combined with oxygen, it forms carbonic acid gas, or the fixed air that fills the bubbles of fermenting beer and bread, and of effervescing soda-water.

8. *Hydrogen* is one of the lightest of gases. It is used to fill balloons, because it is lighter than common air. When it unites with oxygen, it forms water. It is one of the elements of most animal and vegetable substances.

AIR.

9. Atmospheric air is composed of two gases, oxygen and nitrogen, having about twenty-one per cent., or a little more than one fifth of the former, and about seventy-nine per cent., or a little less than four fifths of the latter.

10. *Oxygen* is the most important element in nature. It enters into the composition of all living matter, both animal and vegetable. It is the active ingredient in most acids. United with carbon, it forms carbonic acid; with sulphur, it forms sulphuric acid or oil of vitriol; and with nitrogen, in one proportion it forms

nitric acid or aquafortis, and in another proportion common air.

11. *Nitrogen* is a light, bland gas, and in the air, its main purpose seems to be to dilute the oxygen so that it may not affect the blood and the nervous system too powerfully.

12. Oxygen and carbon have a natural affinity or attraction for each other, so that, when they meet under favoring circumstances, they leave their other combinations and join together, forming carbonic acid.

CHAPTER XIX.

EFFECT OF RESPIRATION ON THE BLOOD AND AIR.

1. *Purification of the Blood.* — THE air, with its oxygen, and the blood, with its waste, composed principally of carbon and hydrogen, meet together in the lungs. The blood absorbs oxygen from the air, and gives back to it both carbonic acid, or oxygen and carbon in a gaseous form, and water, or oxygen and hydrogen, in a state of vapor.

2. The carbonic acid gas and the vapor pass

from the blood-vessels to the air-cells, and the oxygen passes from the air-cells to the blood-vessels, through the thin wall of separation.

3. The blood is thus relieved of its wasted atoms, and is charged with oxygen. By this process, its color is changed from the dark purple of the veins, as seen on the back of the hand, to the bright scarlet of the arteries, as seen when one bleeds from the nose. It is then purified from all its deadly qualities, and is fitted to nourish and give life to the various textures of the body.

4. *Corruption of the Air.* — The air is also changed in respiration, and loses some of its oxygen, and receives some of the atoms of dead flesh, which have been converted into carbonic acid gas and water. It is thus spoiled, in part at least, for the purposes of respiration.

5. The oxygen alone combines with the carbon and hydrogen of the waste atoms, and gives the air its power to purify the blood. About one quarter of this gas in pure air is consumed at each respiration. Air, therefore, when once breathed, has lost a part of its purifying power; and when it has been breathed over several times, it has lost all its means of purifying the blood.

6. *Carbonic Acid carried away.* — The air can hold only a limited quantity of carbonic acid gas. A few respirations of the same air fill it with this gas, and then it can carry away no more. If it is then breathed, the lungs are oppressed; and if then no fresh air can be received into the lungs, suffocation and death follow.

7. *Quantity of Water carried away.* — There is a similar limit to the capacity of the air to take the hydrogen from the blood and carry the watery vapor from the lungs. A definite quantity of this vapor fills or saturates the air, and then it can take away no more.

8. A healthy adult in this climate exhales from his lungs more than thirty ounces of water a day. Every breath carries out some small portion of it. It is invisible in warm weather, because it is so completely dissolved in the air; but in a cold day, it is condensed into a visible cloud of vapor, which comes out of the mouth at every expiration.

9. Air is thus spoiled for the purposes of respiration in three ways — by the loss of its oxygen, so that it cannot purify the blood; and by being saturated with carbonic acid, and filled with vapor, so that it can take away no more from the lungs.

Foul Air in Crowded Rooms.

10. The facts which illustrate these principles are common and familiar. After several persons have been confined for a considerable time in a small and closed room, they begin to feel oppressed and faint; their respiration is laborious and unsatisfactory. It does not give them the relief they want. They suffer from the want of pure air.

11. When we go from the open air into a crowded and unventilated school-room, or hall, that has been long occupied, we perceive at once the foulness of the air, and feel some difficulty in breathing it.

12. This foul condition of the air of any room shows not only that it has lost its purifying power, but that it is also loaded with the dead atoms exhaled from the lungs of the occupants, and is therefore unfit to be breathed again. Consequently, those who live in it cannot be relieved of the waste in their old blood, nor receive the life-giving oxygen for their new.

13. If the elements of those dead atoms,— the carbon of the carbonic acid and the hydrogen of the water,— that should go with the air from the lungs, are allowed to remain in the blood and

F

pass through the arteries again to the system, they produce very serious injury; and if they are allowed to accumulate, they fill the arteries and cause death.

CHAPTER XX.

QUANTITY OF WASTE REMOVED.

1. THE amount of waste atoms that are separated from the textures and need to be carried away, depends upon the rapidity of the changes of the particles that compose the animal body.

2. The rapidity of these changes — the absorption of the old and the deposit of the new atoms — depends upon many circumstances, especially upon the degree of health, the supply of food, and the amount of exercise. They take place more frequently when the body is vigorous than when it is feeble, more when it is well fed than when it is ill supplied with food, and more when the digestion is active, and the textures well nourished, than when chyle is sparingly prepared and the nutriment is meagrely supplied to the general system.

3. The changes of particles are more rapid,

and there are more waste atoms to be carried away, when the body is active than when it is still. Consequently we breathe more frequently, and inhale more oxygen, and exhale more carbon and hydrogen, or more carbonic acid gas and water, when we are at work than when we are at rest; and when we labor violently, as in running, or pumping at a fire-engine, we breathe very rapidly, to carry off the increased waste.

4. The quantity of waste carried out is affected also by the state of the health and spirits, and by the condition of the lungs. More is carried away when the body is vigorous and all the functions are performed with energy, when the body is well fed and well nourished, when the spirits are cheerful, and when life is joyous, than in the opposite conditions.

5. The removal of the waste is especially affected by the condition of the respiratory apparatus. Whatever lessens the size of the chest externally, or fills it internally, and thereby diminishes its cavity, allows less air to reach the blood, to carry off the carbon and hydrogen.

6 Some diseases, such as dropsy of the chest, consumption, lung fever, &c., fill up some part of the space within the chest, which

would otherwise be occupied by the expanding air-cells, and prevent the access of air to the blood. The same effect follows any external compression of the chest which reduces it below its natural size.

7. Whenever less than the due quantity of air is received into the chest, on account either of the smallness of its size or the limited motions of the ribs and diaphragm, or of any internal disease of the lungs, the necessary consequence is the same in all the cases — that less blood is purified, less waste is carried away, and the whole body has a lower degree of life and energy.

Size and Shape of the Chest.

8. The size of the chest is made, by the all-wise Creator, to correspond to the size of the body, in order that the cavity within may receive a quantity of air proportionate to the quantity of blood that needs to be purified, and to the quantity of waste atoms that must be carried away.

9. In the beautiful harmony of the works of God, a large chest is given to a large body, and a small chest to a small body, so that each one may inhale as much air as is necessary to carry

off his own waste atoms, and enjoy that needful freshness of life that only comes from sufficient and free respiration.

10. The natural shape of the chest is conical — large below and smaller above. The lowest two ribs are fixed only at their posterior ends. Their anterior ends float freely in the walls of the chest. The four ribs next above these are attached, by long and somewhat loose cartilages, or bands of gristle, to the breast-bone.

11. This arrangement is intended to allow much more freedom of motion to the lower than to the upper ribs, and a much wider expansion of the chest below than above; it also allows a greater compression of the lower than of the upper part of the chest ; consequently, the lower part, which is made by nature the largest, may, by the artificial pressure of clothing, become the smallest.

MOTIONS OF THE CHEST.

12. The chest expands and contracts like the bellows, and receives at each expansion as much air as its increased cavity can admit. For this purpose, it is necessary that freedom of motion should be a lowed to the ribs, and

room for extension to the abdomen. Whatever binds the body, above or below, restricts the natural motions of the ribs, or of the diaphragm, and prevents the reception of the due quantity of air into the lungs.

Fig. XII.—*Outline of the expanded and contracted Chest.*

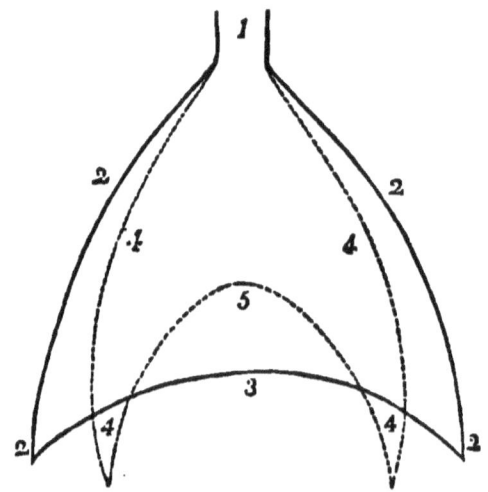

13. In Fig. XII., the dotted lines represent the outline of the sides and bottom of the chest, when it is contracted, and they are drawn in, and the air expelled from the lungs. The full lines represent the same when the chest is expanded, and the lungs are filled with air: 1, the neck; 2, 2, 2, 2, the ribs raised and thrown out; 2, 3, 2, the diaphragm drawn down; 4, 4, 4, 4, the ribs drawn down and

inward; 4, 5, 4, the diaphragm expanded, and its arch thrown up into the chest.

14. The outer cavity is larger than the inner, and it is very plain that the diameter, from 2 to 2, is larger than the diameter from 4 to 4; and that, if any girdle or dress fits close to the smaller cavity, it cannot be expanded to the larger size.

CHAPTER XXI.

AIR NEEDED.—VENTILATION.

1. *Air corrupted.* — A SINGLE respiration consumes about a quarter of the oxygen in the pint of air that is inhaled. The same air may be breathed again; yet, as it is weakened by the loss of a part of its life-giving oxygen, it has not the full power to purify the blood, and cannot carry off the waste as fast as it is deposited in the veins. We therefore need fresh air at every inspiration.

2. *Fresh Air needed.* — Every adult needs, according to the calculations of some philosophers, four cubic feet, and according to the calculations of others, seven cubic feet, of fresh

air every minute, to keep his blood pure, and fit it for the nourishment of his body.

3. When we are out of doors, we obtain the required amount of air without difficulty; but when we are in houses, shops, or halls, the air within these rooms may all be corrupted, and then we need a fresh supply from abroad.

VENTILATION.

4. To meet these wants, and enable the respiration to do its intended work on the blood, every room, shop, hall, and church, every enclosed place where persons live, sleep, work, or assemble, should have some means provided for supplying fresh air to its occupants. Some system of ventilation that will carry the foul air away, and bring pure air in, should therefore be applied to all inhabited rooms, for otherwise the occupants must suffer a depreciation of life.

5. *Effect of impure Air.* — The blood of those who live in crowded rooms is not perfectly purified. They cannot therefore enjoy the vigor and liveliness of body or mind that can only be derived from pure blood. They can neither work with as much power, nor think with as much clearness, nor study with

as much energy, as they could if they had a plentiful supply of pure air to breathe, and of purified blood flowing in their arteries.

ANIMAL HEAT.

CHAPTER XXII.

ANIMAL HEAT.

1. *Animal Heat permanent.* — THE heat of our flesh varies little. A thermometer placed in the mouth, in the coldest days of winter, and in the warmest days of summer, stands at about the same degree — 98°.

2. The temperature of the living body, unlike that of dead bodies, is independent of surrounding objects, and is neither raised nor depressed with them. Our flesh, in winter, is warmer, and, in some days of the summer, is cooler, than the air.

3. Animal heat is not derived from the air, which would rather cool than warm the body,

nor from clothing, which has no active power of warming; but it is dependent on the operations which are performed within the living body. Heat is evolved in the process of removing the dead atoms from their places in the textures.

4. *Heat from Combustion.* — When wood is burned, the oxygen of the air unites with its component elements, — its carbon and its hydrogen, — and forms new compounds — carbonic acid and water. During this process, heat is given out and fire is maintained. This union of oxygen with carbon or hydrogen constitutes what is called *combustion*, and heat is always evolved by it wherever it may take place.

5. *Waste Atoms burned.* — The waste atoms of flesh are composed mostly of carbon and hydrogen. The oxygen which is taken from the air into the lungs, unites with the blood, and passes with it, through the arteries, into the capillaries in every part of the body. In these little blood-vessels, the oxygen meets with the waste atoms, and unites with their carbon and hydrogen in the same manner as it unites with the same elements in wood in the fireplace; and then heat is given out, and the flesh is warmed.

6 *Good Food increases Heat.* — The fire burns the best, and the greatest heat is given out, when there is the best supply of fuel and oxygen. The elements of the waste matters, or the fuel, are originally found in the food, and partake of its character. Consequently, he who is well fed supplies more and better fuel to his internal fire, and is therefore warmer, than he who is ill fed.

7. *Good Air increases Heat.* — This fire is supported by the oxygen of the air. It burns better in pure than in impure air that has lost a part of its oxygen by being breathed once or more. Consequently, persons are warmer when they are supplied with fresh air, than when they breathe foul air.

8. *Exercise and Labor*, which favor the separation of waste atoms, supply more fuel, and increase the fire, and warm the body.

9. *Animal Food* contains more carbon and hydrogen, or more fuel, than vegetable food, and is therefore a more suitable diet for the cold than for the warm season.

10. *Sensations of Heat and Cold.* — Our sensations of heat and cold do not correspond to the degree indicated by the thermometer. When one has been exposed to the air at

zero, and suddenly enters a room heated to 45°, judging by his sensations, he calls it warm. But if another enters the same room from an atmosphere heated to 75°, he calls it cold.

SKIN.

CHAPTER XXIII.

Cuticle.

1. The skin is made to cover over all the internal organs, and protect them from the changes of temperature and from external injury. For these purposes, it can bear contact with outward substances without injury, and endure great variations of heat and cold without suffering.

2. The skin is not a single membrane, but is composed of two layers — the outer, or *cuticle*, and the inner, or the *true skin*.

3. The *cuticle*, or outer skin, is tough and

porous; and, unlike the other textures, it has neither nerves nor blood-vessels, and performs no active functions. It has no sensibility, and feels no pain when it is cut or torn, and does not suffer with the cold and heat.

4. Like the other textures, it is subject to the law of change, or the removal of its old and the deposit of new atoms. But it does not grow by the deposition of new atoms within its substance, and its old atoms are not taken away, one by one, by the absorbent vessels.

5. The new parts grow by the addition of layers to its inner surface from the true skin beneath it; and the old parts are removed or fall off in minute scales from the outer surface.

6. The old particles constitute the branny scurf, which can be easily rubbed off from those parts of the skin which are not frequently washed or exposed to much friction.

7. The thickness of the cuticle differs in the various parts of the body. On the lip it is very thin, soft, and delicate, and on the sole of the foot and the palm of the hand, it is often very thick, hard, and coarse.

8. *Effect of Friction.*— Friction and pres-

sure remove the outer scales as fast as they lose their hold upon the cuticle, and, at the same time, if properly applied, they favor the formation of new layers on its inner surface, and thus increase its thickness.

9. By this means the palms of the laborer and the soles of the barefoot boy become much thicker and harder than the same parts in those who neither work with their hands nor put their bare feet to the ground.

10. *Blisters from sudden Friction.*— But if this friction is applied suddenly and excessively, instead of causing the cuticle to grow thick and hard, it causes pain and inflammation, and raises a blister. In this manner the feet are blistered by walking in very tight shoes, and tender hands are blistered by the use of some tools, or by rowing a boat.

11. Mr. S., who was formerly a farmer, but for several years a book-keeper, undertook, in July, 1847, to mow and rake his hay; but, before he had mowed two hours, he found he had three blisters on his palms.

12. *Corns.*— When pressure is applied for a long period at any particular place, as on the projecting angles of the toe-joints by tight

shoes, the cuticle at these points becomes thickened by the addition of new layers, and thus a *corn* is formed.

13. This thickened cuticle becomes very hard, and causes painful pressure upon the tender skin beneath it. If the external pressure is removed, and loose shoes are worn, and the corn cut away, it is not renewed, and the cuticle remains as thin, and the inner skin as easy, at this point as elsewhere.

Seat of Color.

14. The seat of color is in the inner layer of the cuticle. This differs in different races of men, and in different individuals. It is flesh-colored in the European, black in the African, and copper-colored in the American Indian. It is affected by much exposure to the sun, so that one becomes dark by living in the open air, and pale by confining himself to the shade.

CHAPTER XXIV.

TRUE SKIN.

1. THE inner or *true skin* is a thick and soft membrane lying between the cuticle and the flesh. It is the seat of the sense of touch, and of all the active functions which are performed in the skin. The perspiration and the other cutaneous excretions are prepared in this layer.

2. *Blood-vessels.* — The skin is so abundantly supplied with blood-vessels, that a considerable portion of the blood of the whole body flows in it. When the skin is warm, the blood fills the cutaneous vessels, and gives a lively and rosy hue to some parts of the surface. But cold contracts these vessels, drives the blood from them, and leaves the skin pale.

PERSPIRATION.

3. The perspiratory apparatus consists of the perspiratory glands, (Fig. XIII. 3, 3, 3) which are deeply seated in the inner skin, and the perspiratory tubes, (Fig. XIII. 4, 4,

4,) that pass from the glands, in a winding manner, through the cuticle to the surface of the body.

Fig. XIII. — *Skin and perspiratory Apparatus magnified.*

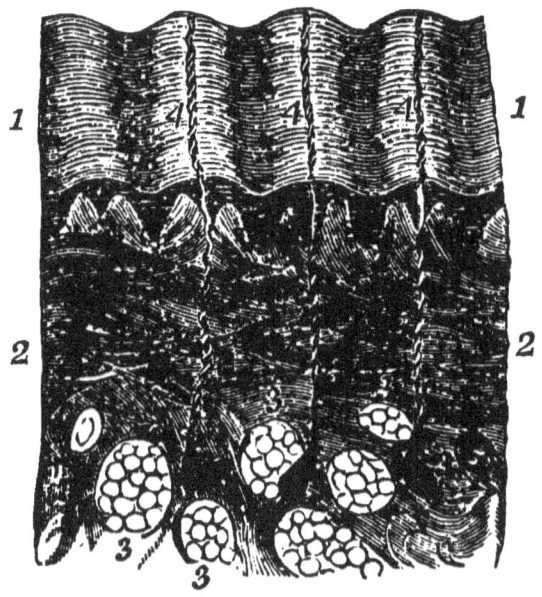

1, 1. Cuticle.   3, 3, 3, 3. Perspiratory glands.
2, 2. Inner skin.   4, 4, 4. Perspiratory tubes.

4. This apparatus performs some of the most important operations in the maintenance of life. It carries off, through the skin of a healthy man, about two pints of fluid a day. This constitutes a large part of the waste of the body.

5. *Insensible Perspiration.* — When the perspiration is visible, and flows in drops, it is called *sweat*. At other times it is invisible, and

forms what is called the *insensible perspiration.* This never ceases. It is incessantly passing off from the inner skin through the cuticle, and its whole quantity is much greater than that of the sensible perspiration.

6. *Quantity of Perspiration.*— When persons are heated by violent labor in a warm atmosphere, the amount of visible perspiration is sometimes very great. Some laborers, who worked at the fire in some of the gas works in London, lost, in three quarters of an hour, different quantities, varying from two pounds eight ounces to four pounds three ounces, by their profuse sweating.

7. *Quantity varies.*—The quantity of perspiration varies with many circumstances. It is more in a warm than in a cold day. It is increased by exercise, by warm drinks, and by certain kinds of medicines.

8. In some diseases the perspiration is abundant, and in others it is suspended. In one stage of fever the skin is dry and parched, and in another it is bathed in sweat.

9. Perspiration is more free, when the skin is in good condition, when it is well bathed and purified of all foul and irritating matters.

10. *Effect of Water-proof Clothing.*—The

perspiration is carried away from the skin by evaporation. If the escape of the vapor is impeded, the flow of the perspired fluid is interrupted. Garments made of water-proof cloth, glazed caps, and India rubber shoes, prevent the passage of the vapor, and therefore interrupt the flow of perspiration, and are on that account unhealthful.

OILY EXCRETION.

11. The skin prepares and sends forth an oily excretion, which gives to the surface its peculiar softness and suppleness. This, like the perspiration, is prepared in the inner layer of this membrane, and sent out through the pores of the cuticle.

12. *Amount of cutaneous Excretion.* —These watery and oily excretions relieve the body of a great proportion of its waste. Sanctorius says, that of every eight pounds which he took into his body, five pounds passed out through the skin.

EFFECT OF CHECKING PERSPIRATION.

13. When a person, in a free perspiration, sits down in a cool breeze, the flow of perspiration may be suspended, and the body cease

to be relieved of its waste, that should go out through the skin; then the flow of the blood through the cutaneous vessels is impeded, and the internal organs are overburdened.

14. When this happens, some internal derangement follows, and then the person is said *to take cold*, which may affect him variously. The respiratory organs may suffer, and he may have catarrh, or lung fever, or pleurisy; his locomotive organs may bear the burden, and then he suffers with rheumatism; or he may have disturbance of the digestive organs, or general fever, in consequence of this interruption of the cutaneous functions.

Effect of Perspiration on Animal Heat.

15. The healthy temperature of the human body is 98°. The internal fire is continually adding to the quantity of heat, and would raise this temperature above the healthy standard, if the surplus were not continually passing off. Part of this passes off, by radiation, from the skin to the air, and part by means of the evaporation of the perspired fluids. These fluids are converted into vapor by the heat of the body, so that one who perspires freely is cooler than others who perspire very little.

16. *Proper Temperature of Rooms.* —When the thermometer, in the air, stands at 98°, or above, the surplus heat is not carried off by radiation from the skin. This temperature is therefore uncomfortable and unhealthful. A temperature of the air considerably lower than that of the body, — as low as 65° or 70° in rooms occupied by sedentary persons, — is the most favorable to health, as well as most agreeable to the feelings.

CHAPTER XXV.

CUTANEOUS ABSORPTION. — SENSIBILITY. — SENSE OF TOUCH.

1. *Absorbent Power of the Skin.* — The skin absorbs, or takes in, some matters from abroad, and carries them into the body. If a particle of matter from a pustule of the smallpox is placed on the skin, it is taken in through the absorbents, and thus the original disease is communicated. In the same manner, the contagion of other diseases, and the poison of ivy, dogwood, &c., are taken into the body.

2. Fluids and other matters are sometimes

received into the body through the *cutaneous absorbents*. In this manner, sailors sometimes quench their thirst by the rain which falls upon them, and persons who were unable to swallow any food have received nutriment sufficient to sustain life for a short time, by being bathed in milk, or other liquid food, which was taken up by the absorbents of the skin.

3. This cutaneous absorption is more active in the night than in the day; and more when the body is feeble and ill fed, than when it is vigorous and well nourished.

4. A man is therefore more liable to be affected by poison and contagion, if exposed to them when he is in a low state of health, than when he is sound and strong, and more when he is hungry, than after he has eaten.

5. Cutaneous absorption is more active when the skin is foul, than when it is cleansed; for the accumulated excretions, which remain on the surface, and the foreign matters that lodge there, stimulate the absorbents to action, and then they take in even this filthy burden.

6. An unclean skin is, therefore, oftener diseased than a clean surface, and cutaneous eruptions are found most commonly among those who seldom bathe themselves.

Cutaneous Sensibility.

7. The inner skin is supplied with more nerves of sensation than the other textures. It has, therefore, exquisite sensibility. The cuticle is insensible; but yet, when it is perfectly cleansed and properly managed, it does not prevent the transmission of impressions through it to the more sensitive membrane beneath.

8. The nerves are unequally distributed in the skin. There are more in the fingers and the lips, and fewer in the sole of the foot or the palm of the hand, than in the other parts of the surface. The sensibility of these parts corresponds to the supply of nerves. The sense of touch is very nice in the finger and the lip, but it is dull in the bottom of the foot.

Sense of Touch.

9. The sense of touch lies in the skin. Its acuteness differs in different persons. It may be cultivated to a high degree. The blind acquire great cutaneous sensibility. They read by applying the ends of their fingers to the surface of the raised letters of their books, and determine, by the touch, the shape and character of each letter, mark, and word. They read,

in this manner, almost as rapidly as others read with their eyes.

10. The draper, who is accustomed to examine and judge of delicate cloths by the sense of feeling, perceives nice distinctions in their qualities, which escape the notice of the bricklayer, whose cutaneous sensibility is blunted by handling rough materials.

CHAPTER XXVI.

CLOTHING.

1. *Necessity of Clothing.* — ALTHOUGH the internal fire develops more heat within the body than is necessary to maintain its healthy temperature, and some of this heat must pass away through the skin, yet more would pass off than can be comfortably spared, in cool climates, if the body were directly exposed to the air. Clothing is therefore necessary to be worn, to protect the body from this excessive loss of heat through the surface.

2. *The Quantity of Clothing* required for health or comfort, depends on the condition of the body, and on external circumstances. The

least clothing is needed when the body is in a state the most favorable to the production of animal heat, when it is the most healthy and vigorous, when it is the best fed and nourished, and in active exercise, and when the spirits are buoyant.

3. The need of clothing depends very much upon habit in regard to dress and exposure. Those who are accustomed to wear much clothing suffer more readily from the cold, and require more garments, than those who usually dress more lightly.

4. Those who live in warm rooms, or work in warm shops, need more protection of garments, when they go abroad, than those who habitually expose themselves to the weather. The shoemaker, when he rides on the stage-coach with the driver, in the winter, must wear more clothing than his companion, or he will suffer more from cold.

MATERIALS OF CLOTHING.

5. The object of clothing being to interrupt the transmission of heat from the body outward, it is effected the best when the garments are made of poor conductors of heat.

6. *Texture.* — Clothes of soft and loose tex-

ture, and with a long nap, are poorer conductors of heat, and are therefore warmer than those which are hard and compact, and worn threadbare. Garments that lie loosely about the body allow less heat to pass through them, and are warmer, than those which fit closely to the skin.

7. *Wool* is the poorest conductor of heat, and makes the warmest garments. Flannel, being woven in loose texture, prevents the effect of sudden cold or chills upon the body, and is therefore proper to be worn next to the skin.

8. *Cotton* is a better conductor of heat, and therefore makes cooler garments than wool, and is better suited for summer than for winter wear.

9. *Silk* and *Linen* are better conductors, and make cooler garments, than either cotton or wool.

10. Linen is much worn for inner garments, in the warm season, because it is cool and pleasant to the skin. But it retains the water of perspiration, and subjects those who perspire freely to sudden chills from evaporation. On this account, cotton is, for many persons, a more appropriate material for summer clothing.

Unclean Garments.

11. All the excretions of the skin — the perspired and the oily matters, and the dead scurf — are first received upon the clothing. The fluid parts of these pass through the garments, when they are sufficiently porous, and are thence carried off in the air. But the more solid particles remain on the garments and soil them.

12. The garments which are soiled by these bodily excretions, become foul and offensive to the sight, the smell, and the sense of feeling. They irritate the skin, and need to be frequently changed and cleansed.

13. *Airing Garments.* — These garments not only need to be frequently washed, but they should never, at any time, be worn for any considerable period without being taken off and aired.

14. The garments of the day should be taken off at night, and those of the night should be taken off in the morning, and neither ever be worn through twenty-four successive hours. When clothes are taken from the body, they should not be laid together in a single pile, but spread, each part separately, and the whole exposed to the air as much as possible.

15. *Airing Beds.*—For the same reason, the bed should not be made up as soon as its occupants leave it, but it should be spread open, and all its parts separated and exposed to the free air, for several hours in the morning.

CHAPTER XXVII.

BATHING.

1. *Foul Skin.*—A CONSIDERABLE portion of the cutaneous excretions, and especially the oily parts, remain on the skin. These mix with the scurfy particles of dead cuticle, and with the dust that floats in the atmosphere, and the whole together form a foul and offensive, and sometimes a glutinous compound, that fills the pores of the cuticle, irritates the inner skin, and interrupts its functions.

2. All these matters should be removed daily from the whole surface, and oftener from the face, hands, and other parts which are exposed to the air, and the dust that floats in it, and to contact with other matters.

3. *The whole Body needs Cleansing.*—

This external cleansing is necessary for the trunk and limbs, which are not exposed, as well as for the hands and face, which are usually uncovered; for all parts of the skin are alike subject to the same law, and are covered and soiled with their own excretions, and need to be freed from this foul burden that is continually thrown upon them.

4. By means of the bath, the pores of the cuticle are kept open, and the excretions are allowed free passage outward, the foul matters are removed, and a cause of irritation is taken away. When this is done, the skin performs its functions with ease and energy, the flow of the blood in the skin is active, the waste is carried off through this outlet freely, and the whole surface is thus made soft and comfortable.

Cold Bath.

5. The cold bath gives the skin more vigor and energy of action than the warm bath. Those who are accustomed to take it can endure the cold of the air better, they need to wear less clothing, and are less liable to suffer from exposure to the changes of the atmosphere, than others who do not bathe.

6. For the perfect health and invigoration of the skin, it is necessary to take the cold bath, not merely occasionally, and in the summer, but daily, and in the winter, continuing and repeating it through the whole year.

7. If the practice of cold bathing is begun in the warm season, and daily repeated and continued through the autumn, into the cold season, the body will gain power of endurance as the cold increases, and suffer but little more from it in the winter than in the summer.

8. Most persons, in good health, feel a pleasant glow upon the surface after the cold bath; but if reaction does not take place, and if the body does not feel warm after the skin has been dried, and proper friction used, then the warm bath must be substituted for the cold.

9. *Bathing benefits all the Organs.*—The advantages of bathing are not limited to the skin; they extend to the internal organs. The lungs, the stomach, the heart, and blood-vessels, the muscles, and the whole nervous system, perform their functions with more energy, and give the whole frame a higher tone of life, when the surface is frequently bathed.

10. *All need Bathing.*— The laborer, therefore, who wishes to exert the utmost muscular

power, and the student, who desires to have the clearest brain, and the idler, who wants the most comfortable bodily sensations, — all accomplish their purposes the most successfully, when they daily bathe themselves in cold water.

11. *Time and Circumstances of a Bath.*— It is the most convenient to take a cold bath when we rise in the morning. But some cannot bathe upon an empty stomach; for, when their bodies are comparatively feeble for want of nutriment, reaction does not take place readily, and they are chilled after their cold bath. And a bath upon a full stomach may suspend or interrupt the process of digestion.

12. The best time for a bath is two to four hours after eating, when the last meal is digested, and the body is refreshed, and before hunger again returns. Yet the robust and healthy may usually bathe at any time excepting immediately after eating.

13. One should not take a cold bath when he is already cold; for then his vital heat is reduced so low, that he cannot safely bear any further reduction.

14. For the same reason, a person is better able to bear exposure to the cold air abroad, if

he warms himself, but does not get into a perspiration, than if he is already cooled before going abroad.

15. *Bathing aids Cutaneous Sensibility.*— The sense of touch is blunted by the accumulation of the cutaneous secretions, and of other foul matters upon the skin; and it is improved by cleanliness and purity. Those, therefore, who wish to make the best use of their sense of touch, as the blind when they read their letters, or as the draper when he examines the textures of his cloths with the fingers, find it necessary to wash frequently.

BONES.

CHAPTER XXVIII.

CHARACTER OF BONES. SKELETON.

1. ALL the organs are connected with a solid framework of bones, which gives to the body its proper form, and supports it in its due position.

Composition of Bones.

2. The bones are composed of lime, or earthy matter, and gelatine, or animal matter, which, when united together, make a very firm, but not a heavy texture.

3. In early years, the bones contain more gelatine than lime; and then they are soft and flexible, and sometimes bend without breaking.

4. In old age, there is more lime than gelatine in the bones; and then they are brittle and very liable to be broken.

5. In the middle periods of life, these two elements are combined in such proportions, that they are neither flexible nor brittle, but have the greatest strength and power to sustain burdens and bear blows.

6. In these several periods of life, the consequences of accidents differ according to the various composition of the bones. In case of a fall, an infant's limbs may bend, an old man's may break, and those of the intermediate age may have power to resist the effect of the blow, and suffer no injury.

7. *Nutrition.* — The bones are supplied with blood-vessels and nerves, and are subject to the law of growth and decay, and to the changes

of substance, by the formation of new and removal of old atoms.

8. *Sensibility.* — In health, they are not sensible, and give no sensation, either of pleasure or pain; but, when diseased, they become very sensitive, and sometimes suffer severely.

9. *Effect of Exercise.* — The bones grow strong by exercise, and weak by inaction. They are, therefore, stronger in the laboring man than in the student; and even the weak bones of the idler may be strengthened by judicious action.

SKELETON.

10. The bones of the head, the trunk, and the extremities, constitute the whole frame, or skeleton. (Fig. XIV.)

11. *The Head*, or *Skull*, seems to be made of a single bone. But it is composed of several bones, which are firmly knit together, with no movable joint between them.

12. *The Trunk* consists of the back-bone, or spinal column, (Fig. XIV. 3, 3,) which supports the upper part of the body, the ribs, (Fig. XIV. 6, 6,) and breast-bone, (Fig. XIV. 5,) which form the chest, and the pelvis, (Fig. XIV. 4, 4,) which extends across from hip to hip, and supports the abdomen.

SKELETON. 115

Fig. XIV.— *The Skeleton.*

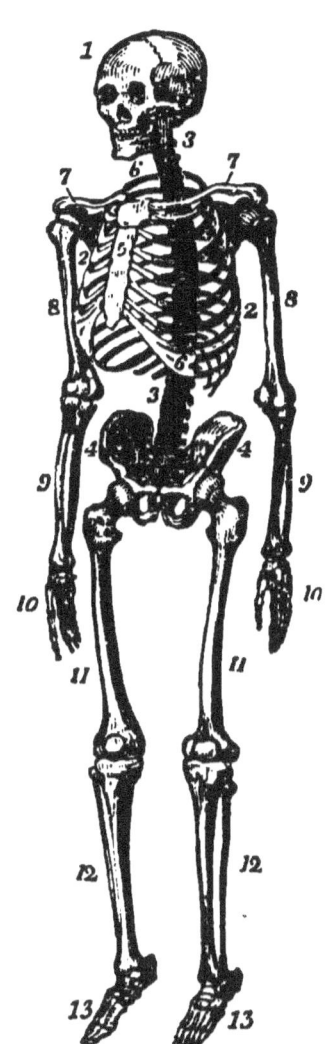

1. Head.
2. Chest.
3, 3. Back-bone.
4, 4. Pelvis.
5. Breast-bone.
6, 6. Ribs.
7, 7. Collar-bones.
8, 8. Upper arms.
9, 9. Fore-arms.
10, 10. Hands.
11, 11. Thigh-bones.
12, 12. Legs.
13 13. Feet.

13. *The upper extremity* (Fig. XIV. 8, 9, 10) includes the shoulder-blade, which lies upon the back of the chest, the collar-bone, which extends from the shoulder to the breast-

bone, the bone of the upper arm, the two bones of the fore-arm, the eight bones of the wrist, the four bones of the hand, and the fifteen bones of the thumb and fingers.

14. *The lower extremity* (Fig. XIV. 11, 12, 13) includes the thigh bone, the knee-pan, the two bones of the leg, the twelve bones of the foot, the two bones of the great-toe, and the three bones in each of the other toes.

CHAPTER XXIX.

SKULL.— SPINE.— CHEST.— HAND.— FOOT.

1. *Skull.* — The round and hollow structure of the skull gives it very great power of

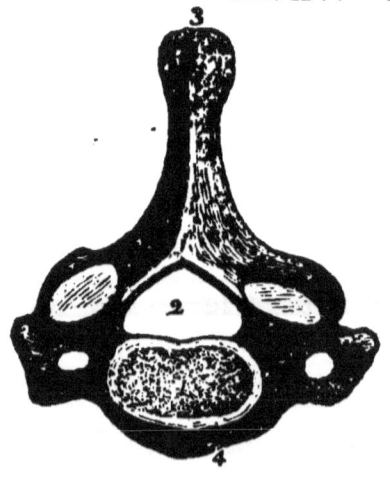

Fig. XV.— *Vertebra.*

1. Body of the vertebra.

2. Hole for the spinal cord.

3. Back part of the bone.

4. Front part of the bone.

BACKBONE.

resistance to blows, and affords the brain a safe lodging-place, where it is seldom injured by the effects of jars, falls, or other accidents.

2. *Spine* and *Vertebræ.* — The back-bone, or *spinal column*, (Fig. XVI.) is composed of twenty-six separate bones, called *vertebræ*. These vertebræ are flat and broad. (Fig. XV.) They are placed one upon another, from the pelvis at the bottom to the head at the top. There is a large hole in the back of each vertebra, (Fig. XV. 2;) and when these bones are arranged together, to form the spinal column, the series of holes in the successive bones forms a continuous channel through the whole length of the spine.

3. *Intervertebral Cartilages.* — Between the several vertebræ are *layers of cartilage*, or gristle, (Fig. XVI. 2, 2)

FIG. XVI. — *Backbone.*

1, 1. Vertebræ.
2, 2, 2. Cartilages.
3. Resting-place for the head.
4. Base of the spine

which vary in thickness from one quarter to three quarters of an inch. They are thickest in the loins.

4. These cartilages are very elastic; they are capable of expansion and compression, like India-rubber, and give the spine its flexibility and freedom of motion.

5. *Form of the Spine.* — The natural form of the spine is straight laterally, but curved forward and backward. (Fig. XVI.) These curvatures are so arranged, that the upper end is vertically above the lower, and the head rests on the top, (Fig. XVI. 3,) directly over the point of support at the bottom. (Fig. XVI. 4.)

FIG. XVII. — *Bones of the Wrist.*

1, 1. Bones of the arm.

2, 2, 2. Bones of the wrist.

3, 3, 3. Bones of the hand.

4. Bone of the thumb.

6. *Chest.* — The chest includes twelve vertebræ, the twenty-four ribs, and the breastbone. (Fig. V. page 59.)

7. *Wrist.* — The eight small bones of the wrist, (Fig. XVII. 2, 2,) are so arranged as to give it great flexibility and strength; yet they are so firmly bound together by ligaments, that they are rarely displaced.

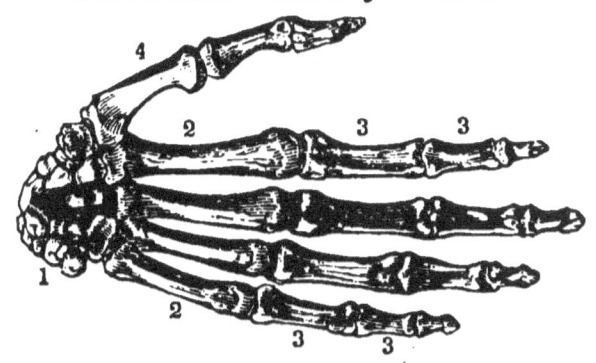

Fig. XVIII. — *Bones of the Hand.*

1. Wrist.   3, 3, 3, 3. Fingers.
2. Hand.   4. Thumb.

8. *Hand.* — The arrangement of the bones of the hand gives it great power and versatility, and makes this organ, with its fingers, a wonderful instrument of usefulness.

9. *Foot.* — The bones of the foot are arranged in the form of a double arch, extending from the heel to the toes, and from side to side, and are bound together by very strong ligaments. (Fig. XX. page 120.)

10. This structure gives the foot elasticity and strength. The bones of the leg rest upon the top of this arch, (Fig. XX. 1,) and thus

the whole frame is supported on this elastic but strong base.

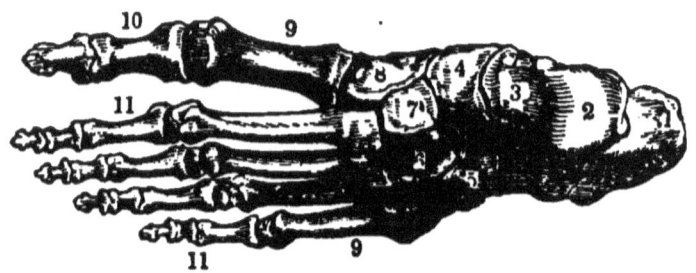

Fig. XIX.—*Bones of the Foot.*

1, 2, 3, 4, 5, 6, 7, 8. Bones of the ankle and instep.
9, 9. Anterior part of the foot.
10. Bone of the great-toe.
11, 11. Bones of the other toes.

11. When we walk, the weight of the body is first received on the heel, which yields

Fig. XX.—*Side View of the Foot.*

1. Bones of the leg.
2. Instep.
3. Toes.
4. Heel.

slightly; it is next partly transferred to the ball of the foot, which yields more than the heel; and, when both are upon the ground, the weight is upon the top of the arch, which yields still more, and thus the force of the shock is so divided, that the body feels no jar.

12. But when a man walks on a ladder, putting the hollow of his foot on the rounds, he derives no advantage from the arch, and feels a jar at every step.

CHAPTER XXX.

JOINTS.

1. THE bones are united by various kinds of joints, which hold them securely in their several places, and yet allow them great freedom of motion.

2. *Hinge Joints.* — The elbow and knee, and the connections of the jaw with the head, and of the parts of the fingers with each other, are merely hinges, or *hinge joints*, which allow the arm, leg, &c., to move in only one line of direction.

3. *Ball and Socket Joints.* — The shoulder-blade has a shallow cup or socket, and the end

of the bone of the arm is rounded like a ball, and is fastened into the socket of the shoulder-blade. The thigh is connected with the pelvis, and the thumb with the wrist, in a similar manner, and these are called *ball and socket joints.* This arrangement gives the shoulder, thigh, and thumb, unlimited range of motion.

4. *Joints of the Spine.* — The connections of the several vertebræ with each other, forming the joints of the back-bone, are merely thick layers of elastic cartilage, which admit of pressure or contraction on one side, and expansion on the other.

5. The head is fixed to the first vertebra by a hinge joint; with this joint we nod and bend the head forward and backward, and this first vertebra is connected with the second vertebra by a hinge like the hook and eye hinge of a gate; with this joint we turn the head round.

6. *Cartilages in the Joints.* — The ends of the bones, in the joints, are faced with cartilage or gristle, that is firm and dense enough to bear the weight that must come upon it, but soft and elastic enough to prevent the effect of jars; yet not so soft as to interfere with the movements of the bones on each other.

Synovial Membrane.

7. These cartilages, and the whole internal surface of the joints, are covered with a smooth lining, called the *synovial membrane*, which prepares and throws into the joint an oily fluid, which keeps the surface of the joints moist and slippery.

8. Sometimes the synovial membrane is diseased, and prepares an unnatural quantity of the fluid, which fills the joints, and causes them to swell. This happens most frequently in the knee, and is often produced there by a blow, or exposure to cold and dampness, as in kneeling on the ground, gardening, &c.

9. *Ligaments of the Joints.* — The bones are held together, at the joints, by very strong ligaments, and the whole covered with a bag or *capsule*, which prevent them from slipping from their places, while they allow all the required range of motions.

10. *Dislocations and Sprains.* — A heavy blow, or fall, may force one bone from its connection with another, and produce a dislocation, by breaking or stretching the ligaments and capsule of the joint. A milder force will

only partially stretch the ligaments, or tear some of their fibres, and produce a sprain.

11. *Strength of the Frame.* — The whole bony framework is thus made very strong, and yet very flexible. The human frame, though perfectly erect, is a series of bones, standing one upon another, and connected, by movable joints, from the head to the feet; and yet, when properly used, it is capable of bearing great burdens.

CHAPTER XXXI.

ATTITUDE.

1. The erect attitude is not only the easiest position for carrying the head, and the body itself, and whatever burdens may be placed upon either, but it is also the most graceful.

2. Whenever the spine is inclined to either side, or bent unnaturally forward, the labor of carrying the head and the trunk is much increased. But when the centre of gravity is placed vertically over the point of support, little exertion is required to keep it in its position.

3. It is easier to support the head, or any weight, on the straight spine, (Fig. XXI. 1) than on the crooked column. (Fig. XXI. 2.)

Fig. XXI.— *Spines Erect and Curved from Side to Side.*

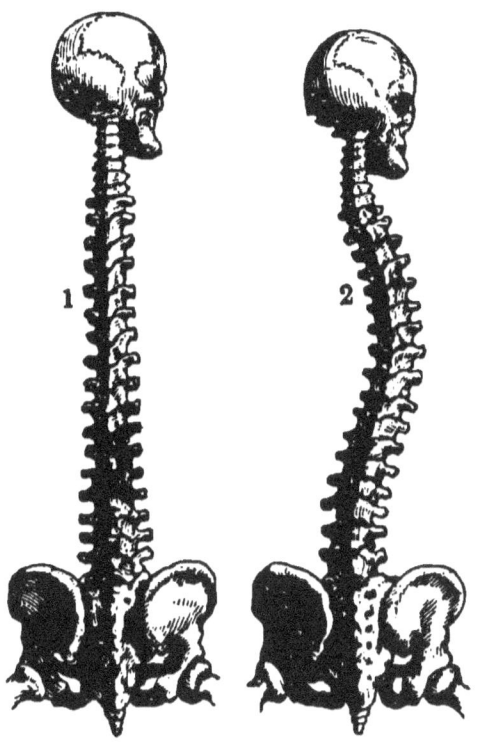

4. This erect attitude gives one a command of his spine, and enables him to carry his burden steadily and securely. Pedlers who carry their merchandise, and servants who carry pails of water, on their heads, have very straight backs; and thus they neither drop their wares nor spill their water.

5. When carried in this manner, the backbone and the lower extremities can bear very heavy weights. Some porters, who have been trained to their employment, can carry more than five hundred pounds on their backs.

Curvature of the Spine.

6. When the form acquires a stooping habit, or is bent from side to side, the attitude and the gait are ungraceful, and the curvature of the spine causes pressure on the spinal marrow, which runs from the brain through the canal in the spine.

7. In consequence of a spinal curvature, the nerves that go from the spinal marrow to supply the internal organs, and the parts below, are disturbed, and extensive, and sometimes serious, derangements of health follow.

8. *Stooping.* — Students bending over their books, watchmakers and engravers leaning over their benches, and seamstresses bending down to their work, are in danger of acquiring a stooping habit, and permanent curvature of the spine.

9. *Lateral Curvature.* — Girls and others, when writing or learning to draw, raise the right shoulder, to lay the right arm on the

table, and let the left arm hang down at the side. In this position, they bend the spine from one side to the other; and, if not counteracted by frequent changes of position, they acquire a lateral curvature of the spine, (Fig. XXI. 2,) which remains permanently upon them, and impairs life ever afterward.

MUSCLES.

CHAPTER XXXII.

ARRANGEMENT AND ACTION OF MUSCLES.

1. *Character.* — THE muscles constitute the lean of meat. They are composed of fibres or strings, which lie parallel with each other, and are generally connected together in bundles. These muscles have the power of shortening themselves, and of drawing their ends toward each other. At the same time, the body of the muscle swells and becomes hard. This

muscular contraction produces all the motions of the animal body.

2. *Attachments.* — The ends of the muscles are generally attached to two different bones. In most cases, one bone is fixed and the other movable; and, when the muscles contract, the movable bone is drawn toward the other, and thus the joints are bent.

3. *Situation.* — The muscles are distributed all over the body, and especially about the limbs. Their main purpose is to bend the joints; yet they are not placed across them; most of them are placed on the body, or part of the limb above the joint, and attached to the bone below by a cord which passes over the joint.

4. *Muscles of the Elbow.* — The muscle that bends the elbow (Fig. XXII. 4) lies on the upper arm, and is attached to its bone (Fig. XXII. 1) near the shoulder; and a cord (Fig. XXII. 6) passes from its lower part, over the elbow, and is attached to the bones of the forearm, a little below the joint, at 6, Fig. XXII. When the elbow bends, the swelling of this muscle, as it contracts, may be easily felt on the inside of the upper arm.

5. The muscle that straightens the elbow

lies on the back of the upper arm, (Fig. XXII. 7:) its upper end is attached to the upper part

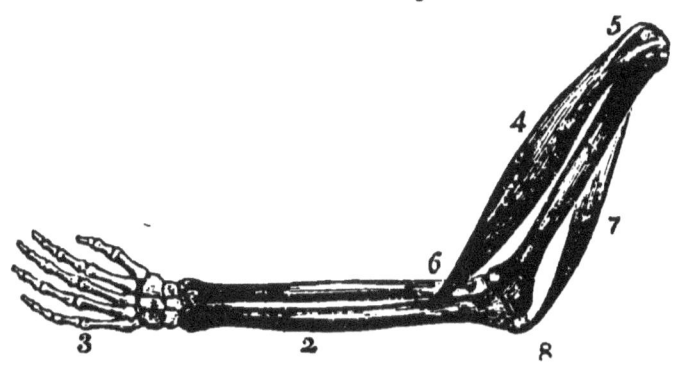

Fig. XXII.—*Muscles of the Elbow.*

1. Bone of the upper arm.
2. Bones of the fore-arm.
3. Hand.
4. Muscle bending the elbow.
5. Its upper attachment.
6. Its tendon, attached to the lower arm.
7. Muscle that straightens the elbow.
8. Attachment to the elbow.

of the bone, and its lower end to the projecting point of the elbow. (Fig. XXII. 8.) The muscles that move the upper arm are placed on the shoulder, back, and chest. Those which move the thigh are on the hips.

6. *Muscles of the Fingers.*—Some of the muscles are placed at a distance from the bone which is to be moved, and the joint which is to be bent. Those which move the fingers are situated on the fore-arm, (Fig. XXIII, 1, 1, 1, 1,) and long tendons, or cords, (2, 2, 2, 2,) pass over the wrist and hand to the remote fingers.

I

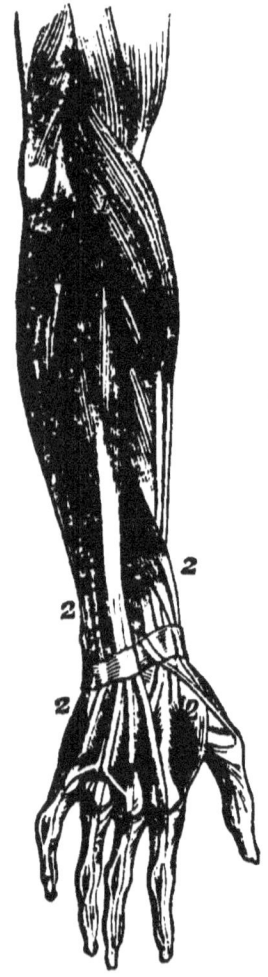

Fig. XXIII.—*Muscles of the Fingers.*

The swelling of these muscles can be felt on the inside of the arm when we bend the fingers, and on the outside when we straighten them.

7. The muscles that move the toes lie on the leg above the ankle. Long tendons run from them over and behind the ankle joint, and over and under the foot to the distant toes.

8. *Form of Muscles.*— The form of the muscles differs in various places. Most of them are long and roundish, as in the fore-arm, (Fig. XXIII. 1, 1, 1,) others are flat, as on the abdomen and breast; some are fan-shaped; some are in the form of a ring, as those which wind around, and close the eyes, and mouth, and the circular fibres of the muscular coat of the stomach. (Chap. IV. § 6.) The heart is a hollow muscle, or muscular bag.

9. *Action of the Muscles.* — The muscles are arranged to perform all the motions that are required. The hinge joints, the elbow, knee, &c., are provided with two sets of muscles, (Fig. XXII. 4, 7,) one to bend, and one to straighten the limb. The other joints are supplied with more muscles, according to the variety of their motions. The shoulder has one set to lift it up, and another set to pull it down; one to draw it forward, and another to draw it backward; and also other muscles to move it in other directions.

10. *Voluntary and Involuntary Muscles.* — There are two classes of muscles, the voluntary and the involuntary. The voluntary are under the control of the will, and include the external muscles situated on the limbs, back, &c. These act only when we direct them. The involuntary include the heart, and the muscular coats of the digestive organs, over which we have no control. They act when we are asleep as well as when we are awake. Other muscles, as those of respiration, are involuntary in as far as they act without direction, even in sleep; and also voluntary, inasmuch as we may quicken or temporarily suspend their actions.

CHAPTER XXXIII.

MUSCULAR ACTION.— EXERCISE.

1. *Control of the Will.* — The voluntary muscles act with great precision. When the mind determines any motion, the muscles which move the limb, contract just to the degree sufficient to produce the motion that is required. Each muscle performs its own part, and no more, to produce the intended result.

2. If we will to lay the right hand on the head, a muscle on the top of the shoulder lifts the arm to the proper height, and another draws it forward to the proper direction; a muscle on the upper arm bends the elbow, and another on the fore-arm bends the wrist, and all these combined together lay the hand on the head.

3. *Coöperation of Muscles.* — All the motions are produced by the action of single muscles, or by the combination of several. In some actions, several sets of muscles coöperate. In walking, those on the front of the hip, and on the back of the thigh, and the bending or *flexor* muscles of the foot, of one side, contract and bend the hip, knee, and ankle, and

raise that limb, while the muscles on the back of the hip, the front of the thigh, and the back of the leg, of the other side, straighten the other hip, knee, and ankle, and place that foot on the ground.

4. Not only the motions of the limbs and of the bones on each other, but all the movements of the flesh, are produced by muscular action.

Fig. XXIV. — *Some of the Muscles of the Face.*

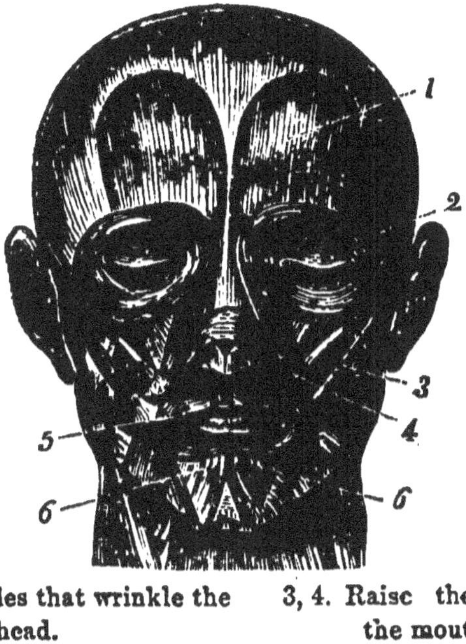

1. Muscles that wrinkle the forehead.
2. Closes the eye.
3, 4. Raise the corners of the mouth in smiling
5. Closes the mouth.
6, 6. Draw the lower lip downward.

By means of this power, we laugh and talk, we roll the eyes, and create the various expressions of the countenance, the blood is circulated, and the contents of the digestive organs are carried onward.

5. *Muscles grow by Use.*—Muscles grow large and strong by use. The laborer has larger and stronger muscles than the sedentary lady. This growth and strengthening are limited to those muscles which are active. The sailor has strong and brawny arms, and comparatively little and weak legs, while the rope-dancer has large legs and small arms, and the farmer has large muscles on the whole of his trunk, and on all his limbs.

Exercise.

6. *Need of Exercise.*—Some muscular action is necessary, not only for the growth and the strengthening of the muscles themselves, but for the health and strength of all the other organs. Exercise quickens the respiration and the circulation of the blood; it favors digestion, perspiration, and the removal of the waste, and aids the action of the brain.

7. *Effect of Exercise on Digestion.*— As the waste is increased by muscular action, there is need of more new atoms, and a demand for more food; and with these comes an increase of

digestive power. Active persons seldom suffer from want of appetite, or from indigestion, which frequently trouble the sedentary and the idle.

8. If a man's occupation is not sufficiently active to give him as much muscular action as his health requires, it is then necessary that he leave his quiet employment, for some part of each day, and take active exercise abroad.

9. By this means students acquire a greater command of the brain, and study with more effect; sedentary mechanics gain greater control of their muscles, and direct their hands and use their tools with more precision; and they all gain in energy of life, in clearness of mind, and cheerfulness of spirit.

10. *Amount of Exercise.* — The exercise cannot be alike for all. One man is benefited by a degree and amount of muscular action which would exhaust and weaken another. A man's strength must be the standard by which he may measure the amount of exercise that is proper for him. It must be active in proportion to his bodily vigor, and protracted in proportion to his power of endurance.

11. No universal rule can be given for the amount of exercise which is necessary. But those who enjoy good health, and who wish to

exercise only sufficiently to maintain it, should walk or work about three hours each day.

12. *Kinds of Exercise.* — Agricultural and horticultural employments call all the muscles into action, and give energy to the whole frame, and are therefore the best kinds of exercise. Walking is good, and is within the reach of every one; and if working with tools is added, the whole frame will be exercised, and the best effects produced.

13. *Conditions of Exercise.* — When the body is weak from long fasting, muscular action rather exhausts than increases its vigor. Exercise immediately after eating suspends or retards the digestive process, and postpones nutrition, (Chap. IX. § 10.) Exercise just before eating, and especially if it is fatiguing, exhausts much of the energy that should be reserved for the digestion of the coming meal. (Chap. IX. § 11.) Sedentary men, therefore, should not take their exercise when they are hungry, nor just after their meals.

CHAPTER XXXIV.

LABOR.

1. WHEN the body is in good health, the muscular system can bear much more action than the mere maintenance of health requires. This muscular action may be increased, and at the same time add to the strength of the muscles.

2. *Limit to the Power of Labor.*—But there is a limit to this power of labor, which cannot be passed without exhaustion and weakness. If this limit is passed, and this excess of action is frequently and continually repeated, the body becomes permanently weak, and premature old age is induced.

3. In order to maintain the muscular strength undiminished, the laborer must exhaust no more power during the day than can be restored in the night. By this judicious regulation of exercise, the body may be fatigued at evening, but it is completely refreshed and renewed every morning.

4. *Excessive Labor.* — If the muscular action is thus managed, and the labor is thus

12 *

arranged, the body may retain its full power of exertion to a late old age. But the strength of most laboring men begins to fail long before the natural term, in consequence of the excess of their muscular actions, beyond their daily power, during the early and middle periods of life.

5. Although these men, who toil thus excessively, may seem to gain by their great labors while their strength lasts, yet this gain is more than lost by their premature weakness, and the necessity of slackening in their exertions before the natural period of old age would invite them to rest.

6. *Action and Rest.* — The muscular system needs alternate action and rest. Varied labor, such as gardening, &c., which exercises many and various muscles successively, and allows them intervals of rest, is less fatiguing than that which exercises a few muscles for a long period without intermission, as sawing wood, &c.

7. *Violent Action.* — Very great exertions are sometimes followed by permanent injury. Some, who are not accustomed to hard labor, are injured by pumping with all their energy at a fire-engine, or engaging in a boat-race.

8. *Labor of Youth.* — In early years, before

the bones become consolidated, and while the muscles are weak, the exercise may be active, but never violent; the labor must be light, and never long continued. Children and youth have not much power of endurance, and if hard labor is required of them, it will prevent the future growth and strengthening of their bones and muscles.

CONDITIONS OF LABOR.

9. *Day and Night Labor.* — The light and air of the day are more favorable to muscular exertion than the darkness and air of the night, and the laborer can accomplish more work in an hour, and is less exhausted by it, while the sun is above, than while it is below, the horizon.

10. The workmen are sooner worn out in those occupations that require the labor of the night, than in those which require the labor of the day. Those who work only a part of the night, and rest as much of the day, suffer in proportion to their violation of the natural law of exercise.

11. *Food of the Laborer.* — Labor creates great waste of the atoms of flesh, and great demand for new atoms to supply their places. The laborer, therefore, must have a sufficient

quantity of good and nutritious food, that can easily be digested and converted into blood and flesh.

12. *Digestion and Circulation.* — For the same reason, it is necessary that the laborer's digestive organs be in sound condition to convert his food into chyle, and fit it for the blood, and that his heart and arteries be able to carry this blood to the textures, and that his lungs be in good health to carry off the waste matter.

13. *Respiration.* — Persons suffering from consumption or other diseases of the lungs, cannot inhale air sufficient to carry off the waste caused by great labor; and if they attempt to run, or exercise violently, they breathe more rapidly, to bring more air to the blood, and often their respiration is very quick and distressing.

14. For the same reason, a close dress, or any thing that contracts the size, or prevents the free expansion of the chest, interrupts the removal of the waste atoms, and lessens the energy of muscular action, and the freedom of labor.

15. The healthy state of the digestive and circulatory system, which nourish the textures, and of the respiratory and cutaneous systems,

which carry off the waste atoms, is necessary for the strength of the muscles and for their power of action. But this action may be quickened or impeded by the state of the nervous system, or by matters which affect it.

16. *Effect of Spirit on Labor.* — Spirit quickens the vital actions, but it does not add to the strength of the body; it only gives one a greater command of the power that is already in him, and enables him for a moment to do more work: but this extraordinary and unnatural exertion is made at the cost of future power, for he is sooner exhausted by it.

17. *Effect of Cheerfulness.* — A better stimulant is in the state of the mind and feelings. Cheerfulness favors muscular action, and increases the power for labor; but melancholy impedes both. A man can work more effectually when the mind approves, and the heart is satisfied with, his object and progress, than when he is discontented and hopeless.

BRAIN AND NERVES.

CHAPTER XXXV.

BRAIN AND NERVES.

1. *Head.* — The skull is a hollow box composed of several bones united so firmly together as to bear heavy blows without breaking or allowing injury to the brain within it. (Chap. XXVIII., §11, p. 114.)

2. *Brain.* — The brain is a large collection of nervous matter, which is placed in the skull, and fills all its cavity. Its texture is soft and very delicate, and may be broken with the fingers. It is covered and held together by three membranes.

3. *Coverings.* — The outer covering of the brain is thick and very strong, and supports the organ. The two inner coverings are very soft and delicate, and give it an easy cushion to rest upon. These prevent its contact with the hard bone, and protect it from the effect of accidents, blows, jars, &c., that otherwise might injure it.

4. *Blood-vessels.* — The brain is supplied with many arteries and veins, and a large amount of blood flows through it. It is sub-

Fig. XXV. — *Brain and Spinal Cord.*

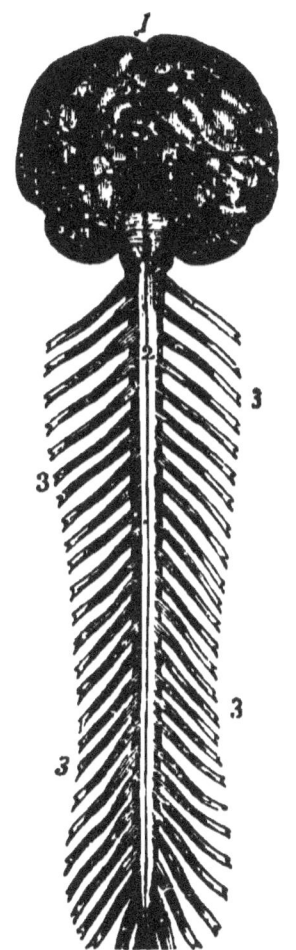

1. Brain.   2. Spinal Cord.   3, 3, 3, 3. Spinal Nerves.

ject to the law of growth and decay. Its old atoms are exhausted and die, and then are carried away by the veins, and new atoms from the arteries take their places.

Spinal Cord.

5. The spinal cord (Fig. XXV. 2, p. 143) is a large nerve, or bundle of nerves, extending from the base of the brain, (Fig. XXV. 1,) through the whole length of the back-bone, in the channel formed by the series of rings in the vertebræ.

Nerves.

6. The brain and spinal cord are connected with the whole body by means of nerves, or strings of nervous matter that extend from the brain and spinal cord to the various parts of the body.

7. Twelve pairs of these nerves extend from the brain to the organs of sense, and to other parts of the body. Thirty pairs extend from the sides of the spinal cord, (Fig. XXV. 3, 3, 3, 3,) through the spaces between the vertebræ, and the holes in the lowest bone of the spine, to the respiratory, digestive, and circulatory organs, and to the muscles and bones of the body and limbs.

CHAPTER XXXVI.

USES OF THE NERVES.

1. *Nerves of Sensation and Motion.*— There are two classes of nerves, that extend from the spinal cord to all parts of the body,— one called the *sensory* nerves, or nerves of sensation, and the other called the *motory* nerves, or nerves of motion. The sensory nerves pass from the posterior, (Fig. XXVI. 2,) and the motory from the anterior, part of the cord. (Fig. XXVI. 1.) These are united in pairs in one sheath, when they first come from the

Fig. XXVI. — *Section of the Spinal Cord, and Roots of the Spinal Nerves.*

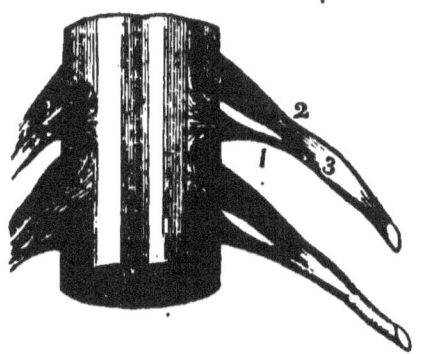

1. Anterior or motory nerve. 2. Posterior or sensory nerve.
3. Union of both nerves.

back-bone, but they are soon separated, and pass to different points.

2. The brain may be considered as the origin and centre of all motion and sensibility; and the nerves convey these powers to the several organs and parts of the body. The muscles depend on their connection with the brain for their power of action, and all the parts of the body depend upon it for their power of feeling.

3. The nerves of sensation convey to the body the power of feeling. The nerves of motion convey the power of motion. All the parts and organs have, through these nerves, communication with the brain, which is necessary for their life and action.

4. *Nerves of Communication necessary.* — If the sensory nerve is cut off, or if the communication of any part with the brain is interrupted by pressure on the nerve, or otherwise, the part loses its sensibility, and cannot feel. If the motory nerve is cut off or pressed, the part cannot move.

5. *Action of Sensory Nerves.* — When impressions are made upon the distant and outer terminations of the sensory nerves in the eye, or ear, or the flesh in the various parts of the

body, they are carried along the sensory nerves to the brain, where sensation is excited. But, if this communication is interrupted by disease or injury of the nerves, there is no sensibility in the outer part, and then the eye is blind, and the flesh has no power of feeling.

6. *Action of Motory Nerves.* — The impulse of the will is carried from the brain, along the motory nerve, to the muscles; but, if this nerve is injured, the muscles, in which it terminates, receive no command from the will, and have no power of action, and the limb is palsied.

7. *Effect of Pressure upon the Nerves.* — Palsy and insensibility are produced, also, when the nervous communication with the brain is interrupted by pressure. Thus, the foot loses its power of motion and feeling when the nerve that connects it with the brain is pressed, by crossing the legs, or otherwise, and then the foot is said to be asleep.

8. In some disorders of the motory nerves the muscles in which these nerves terminate may be stimulated to uncontrollable action. In disorder of the sensory nerves, there may be pain, as in tic douloureux. When these nerves are injured, there may be a tingling sensation; as, when we strike the elbow and hit the nerve,

we feel a sharp tingling in the fingers, where the injured nerve terminates.

9. The nerves of special sense — of the eye, ear, &c. — may be excited by disease or irritation, and carry impressions to, and create their peculiar sensations in, the brain. Thus, when the optic nerve is disordered, or jarred, one sees flashes, and when the auditory nerve is diseased, he has ringing in his ears.

10. *Internal Organs connected with the Brain.* — The nervous communication with the brain is necessary for the internal organs, to enable them to perform their peculiar functions. The stomach, liver, lungs, heart, &c., all have their nerves; and, if these are cut off, they lose the power of action, and then the stomach secretes no gastric juice, and digests no food, the liver prepares no bile, and the lungs fail to purify the blood.

11 For these several functions, it is necessary, not only that the organs in which they are performed should be in good health, but that the brain, and the nerves that connect them with it, should also be in good condition.

12. *Sympathy between the Brain and Organs.* — There is a mutual sympathy between the brain and the other organs. When both

are well, and perform their operations easily, we are not conscious of this connection; but when either is deranged, we feel the disturbance in both.

13. When the brain is pressed with blood, the muscles are paralyzed, and the lungs breathe slowly and laboriously. When the brain is jarred, the stomach vomits. When the stomach is nauseated, or the liver is torpid, or the lungs do not purify the blood, the brain is distressed, and the head aches. And this organ becomes feeble and indisposed to action when the organs of nutrition fail.

14. *Effect of Action on the Brain.* — The brain needs exercise and rest; it gains power by action, it is exhausted by excessive labor, and it becomes weak by inactivity.

CHAPTER XXXVII.

BRAIN AND MIND.

1. THE brain is the seat of the mind and of the affections, and the organ through which all our moral and mental powers are manifested. It is therefore necessary, for the free and pei-

fect operation of these powers, that the brain should be in good condition.

2. *Connection of the Mind and Brain.* — When the brain is vigorous, the mind acts with energy. When a person's head aches, he is unfitted for study, and then his mind is confused, and will not labor, because its organ is indisposed to action. When the brain is excited by spirit, the mind acts with energy, and the ideas flow rapidly; but when it is pressed with blood, as in apoplexy, the mind is torpid.

3. When the brain is diseased, it acts irregularly, and then the mind is deranged, or delirious, and has strange thoughts, or false ideas, or the feelings may be perverted, and the temper unnaturally irritable, suspicious, melancholy, or exhilarated.

4. While the body lives, the mind is inseparably connected with the brain, and is therefore limited, in its actions, to the power of this organ. When the brain is torpid, the mind is in the same dull condition; and when this physical organ is weary, the mind wants rest.

5. *Connection of Mind and Body.* — As the brain is affected by the state of the other organs, the mind obeys the same law, and also

suffers with them. Sickness of the stomach renders the mind powerless. When a man has overloaded his stomach, he is indisposed to study or think. Dyspepsia often creates an irritable temper, and a torpid liver sometimes produces a melancholy spirit. When the student takes no exercise, his mind is comparatively feeble; and when one is fatigued with labor, he is averse to mental action.

6. *Effect of the Mind on the Body.* — The mind, on the contrary, affects the various organs and their operations. Cheerfulness and mental calmness favor digestion, and melancholy and an over-wrought mind promote dyspepsia. The muscles act with more energy, and the laborer works with more effect, when his mind is contented; but, when the mind is dissatisfied, or the spirit is oppressed, the limbs are languid, and labor is wearisome.

7. Excessive mental action impairs the whole body and its organs. The laborious student wastes his digestive powers, the overanxious merchant loses his flesh, the gloomy hypochondriac loses his muscular strength, and despair aggravates any bodily disease.

8. In view of the mutual influences of the mind and body, the perfect health and action

of either is necessary for the best health and action of the other. As calmness of mind, cheerfulness, and serenity, favor the operations of the bodily organs, so disturbance of these organs may cause the mind to falter, the spirits to droop, or the temper to be ruffled.

CHAPTER XXXVIII.

THE EYE.

1. The eyes are placed in deep, bony sockets, which are lined with layers of fatty matter that serves as a soft bed for the eyeball to roll in.

2. *Composition.* — The eye is composed of three transparent substances — the *aqueous* and the *vitreous humors*, and the *crystalline lens*. The aqueous or watery humor is in front, (Fig. XXVII. 2;) the crystalline lens is of somewhat dense texture, and lies next behind the aqueous humor. (Fig. XXVII. 3.) The vitreous humor is of a pulpy consistence, and constitutes the greatest part of the eyeball. (Fig. XXVII. 4.) The whole globe is covered by strong membranes. (Fig. XXVII. 1.)

3. *Lids.* — The lids cover the eye in front. The circular muscle, (Fig. XXIV. 2, p. 133,)

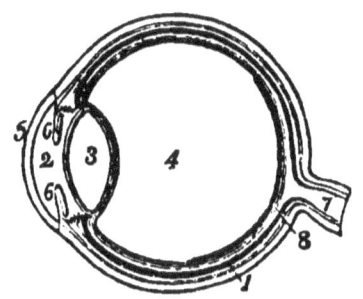

Fig. XXVII. — *Section of the Eye.*

1. Coverings.
2. Aqueous humor.
3. Crystalline lens.
4. Vitreous humor.
5. Cornea.
6, 6. Iris.
7. Optic nerve.
8. Retina.

which closes them, sometimes acts involuntarily, and then we wink unconsciously, to protect the eye from injury.

4. *Cornea and Pupil.* — The transparent cornea is in the front of the eye, (Fig. XXVII. 5,) and the pupil is the small, dark circle in the centre. The iris surrounds the pupil, and regulates its size, to admit more or less light.

5. While we are in a light room, the pupil is contracted; going suddenly into the darkness, we see with difficulty, until the pupil is enlarged and admits more rays. Going suddenly from a dark to a light room, the enlarged

pupil admits so many rays, that we are dazzled; but soon the pupil contracts, and then the abundant light ceases to disturb the eye.

6. *Tears.* — The tears are prepared in the lachrymal gland, at the upper and outer corner of the socket, and flow, over the eye, to the inner corner, and pass through the lachrymal duct into the nose. They keep the eye always moist; and, when any foreign body gets into it, they flow abundantly, to carry it away. Their flow is increased, also, with some moral excitements, as grief, &c.

7. *Sympathies.* — The eye has sympathies with the other organs, and suffers with them. Motes in the eye, and some species of blindness, and other diseases of this organ, are sometimes caused by diseases of the digestive organs, by want of food, and by bad air.

8. *Bathing.* — The eye suffers from uncleanness, and needs, therefore, to be bathed every morning, and at other times, especially when exposed to dust.

Vision.

9. *Optic Nerve.* — *Retina.* — The nerve of sight, or *optic nerve,* extends from the brain into the back part of the eye, (Fig. XXVII. 7,)

where it is expanded, and forms what is called the *retina.* (Fig. XXVII. 8.) The rays of light from any object form its image on the retina. The impression of this image is then carried along the optic nerve to the brain, where the sensation of vision is created.

10. Vision is improved by exercise, and grows dull by neglect. The eye is wearied with labor, and wants rest and sleep, and it is exhausted by over-exertion. It requires sufficient light for its action, and is injured by being used for reading, sewing, &c., in any feeble light.

11. *Near-sightedness.* — The eye accommodates itself to receive impressions from objects both near and remote. But if it is long used to see near objects exclusively, it loses the power of adapting itself to see distant things, and thus students, proof-correctors, &c., become near-sighted.

12. *Far-sightedness.* — The eye gains great power of adapting itself to remote objects, when much used to look at them; and thus the sailor and the traveller become far-sighted. In old age, the eye loses its power of seeing near objects, and we then need the aid of convex glasses to enable us to read.

CONCLUSION.

1. THE preceding descriptions of the organs of the human body and their uses show how fearfully and wonderfully we are made, how nicely are all the works of the benevolent Creator fitted to their intended purposes, and how beautiful are the harmonies between the organs and powers of the living body and external matter.

2. We also see that the material body is not left to its own guidance, but to each one is given a mind that shall direct it. Every human being is thus made responsible for the care of his own health, and the preservation of his own life.

QUESTIONS.

INTRODUCTION.

CHAPTER I. 1. What is anatomy? 2. What is physiology? 3. Of what use is a knowledge of these subjects? 4. Why is this necessary? 5. In what ways are we appointed to take care of our bodies? 6. What have we to do with regard to our food? 7. What for the lungs? 8. What for the skin? 9. What for the muscles? 10. What for the mind? 11. Why govern the passions and appetites? 12. 13. What knowledge is needed to fulfil these responsibilities?

DIGESTION AND FOOD.

CHAPTER II. 1. What is the consequence of want of food? 2. What changes does food undergo? 3. What is the object of food? 4. What changes are going on in the body? How does food aid them? 5. By what means is food converted into flesh?

6. What constitute the digestive organs? 7. What are included in the mouth? 8. How many teeth in the mouth? What kinds? How situated? 9. Of what are they composed? Describe the enamel. 10. What causes decay of the teeth? 11. What causes the toothache?

CHAPTER III. 1. What moistens the mouth? How? 2. How is the saliva made to flow? 3. What are its uses and abuses? 4. What is the use of the mouth? The teeth? The saliva? 5. How is the food prepared for the stomach? What moistening is needed?

6. What is the œsophagus? Where situated? 7. What fibres compose it? How do they act? 8. Describe the process of swallowing

CHAPTER IV. 1. Where is the stomach? 2. What is its size? 3. Of what is it composed? 4. What are the offices of the peritoneal coat?
5. Describe the muscular coat. What power have these fibres? 6. How are they arranged? 7. How do they act? 8. What is the object of the mucous coat? 9. How does tripe illustrate the human stomach? 10. What are the offices of each coat?

CHAPTER V. 1. What is the gastric juice? Where and how prepared? 2. When is it in the stomach? How does it come? 3. How does its flow resemble that of the saliva? 4. What does Dr. Beaumont say of the gastric juice? 5. How much food can the healthy stomach digest? What if more is eaten? 6. What causes the sensation of hunger? 7. When does it begin? And when cease? 8. What guide have we for the quantity of food?

CHAPTER VI. 1. What is the state of the muscular coat after eating? 2. How is the food in the stomach kept in motion? 3. Describe the relative situations of the heart, lungs, diaphragm, and stomach. 4. 5. How does the respiratory process affect the stomach? 6. What is chyme? 7. What is done with the digested food? 8. 9. 10. What is the pyloric valve? What its use? Is it always true to its duty? 11. 12. How long does the digestive process last?

CHAPTER VII. 1. Is food always digested in the same time? Why? 2. Is the time the same in the same person? 3. What means of observation had Dr. Beaumont? 4. 5. In what time are some common articles of food digested? 6. Why is not this a certain rule?
7. What effect has drink with food? 8. What is Dr. Warren's opinion?
9. Why cannot the stomach act on food that is imperfectly chewed? 10. How does gastric juice act on food? 11. Does

it act with equal ease on all kinds? 12. What becomes of indigestible food?

CHAPTER VIII. 1. Describe the alimentary canal. 2. The peritoneal coat. 3. The muscular coat. 4. The mucous coat.
5. Describe the tubes in the mucous coat. 6. The lacteal system. The lacteal absorbents.
7. What change in the chyme after entering the canal? What is chyle? 8. How does the innutritious part of the food affect the body if allowed to remain? 9. Is the quantity of chyle the same in all chyme? Why not? 10. 11. What becomes of the chyle?
12. What are the several processes of digestion? 13. How does imperfect mastication affect nutrition?

CHAPTER IX. 1. What is man's duty with regard to food? 2. To what must the supply have reference?
3. How does appetite aid in this matter? 4. What is a false appetite? 5. How is the propriety of taking food manifested? 6. What is the consequence of eating when not hungry? 7. What guide have we?
8. What circumstances affect the necessity of food? Why should the laborer eat much? 9. What others should eat much? Why? 10. Why is rest after eating necessary? 11. Why before? 12. For whom are these cautions most necessary? Why?

CHAPTER X. 1. Do all kinds of food affect the body alike? 2. What are the properties of meat? When and by whom should it be eaten most freely? 3. How would the same food affect people in opposite climates? 4. Why do different constitutions require different food?
5. What is the effect of condiments and stimulants? 6. What is the difference between a natural and a perverted taste? 7. How far and when should the taste be consulted? 8. Why can there be no rule of diet for all men?
9. How soon after a meal should we eat again, and why?

10. Are the usual meals proper? and why are the intervals unequal? 11. When should the breakfast, dinner, and supper, be taken? 12. What is the effect of late suppers?

CIRCULATION OF THE BLOOD.

CHAPTER XI. 1. What is the sole object of food? 2. How does the chyle do this? 3. Of what does the apparatus of circulation consist? 4. What is the office of the heart? 5. What are the course and office of the arteries? 6. Of what use are the capillaries? 7. Of what the veins? 8. Describe the heart. 9. Into how many parts is it divided? Where and how do they communicate? 10. What are the valves? How do they operate? 11. What other valves?

CHAPTER XII. 1. What vessels are connected with the heart, and how? 2. How are they divided, arranged, and distributed? 3. 4. How are the arteries and veins said to be related to the heart and body? 5. Where do the arteries and veins nearly meet? 6. Describe the general circulation. 7. The pulmonary circulation. 8. The double circulation. 9. How often does the heart beat, and how much blood is sent out? 10. What is pulsation? 11. What affect this action of the heart? 12. How much blood is in the body, and how often does it pass through the heart? How much passes through it in a day?

NUTRITION.

CHAPTER XIII. 1. Why does the blood flow through the body, and what is done with it in early years? 2. What during the whole of life? 3. What becomes of the dead atoms, and how are their places filled? 4. How does this change affect the body? 5. What are the two necessities for food? 6. What parts are formed from the same blood? What offices do the capillaries perform? 7. How do the nourishing vessels and absorbents coöperate? How do they coöperate in various ages? 8. 9. What is the effect of exercise on changes of particles? How does this affect the necessity of food? 10. How are nutrition and absorption affected by age, and how by exercise? 11. What are the different states of the blood in the arteries and veins? What their color? 12. How are the exhausted and dead atoms of the body disposed of? 13. What quantity carried out each day? What are the consequences if they remain in the body?

RESPIRATION.

CHAPTER XIV. 1. Of what does the blood consist in the right side of the heart? 2. Why is it unfit to nourish the body? 3. How can it be made good again? 4. Where are the lungs situated, and with what are they furnished? 5. Describe the chest. 6. What are the bones of the chest? 7. How many ribs on each side, and how are their ends attached? 8. What motions have they? 9. What is their course? 10. What advantage from this oblique position? 11. How is the size of the chest enlarged? 12. Explain Fig. VIII.

CHAPTER XV. 1. What move the ribs? 2. Where are these muscles? 3. How do they act? 4. Explain the combined action of the intercostal and spinal muscles. 5. What contracts the chest? 6. What is the diaphragm? 7. To what are its edges attached? Where is its centre? 8. What is its position in action and at rest? 9. How are the mechanical parts of inspiration performed? 10. Does the diaphragm act long? 11. Where are the diaphragm and the abdominal muscles placed? 12. How does the diaphragm act on the digestive organs and abdomen? 13. When and how do the abdominal muscles act, and with what effect? 14. How is the air forced out of the lungs?

CHAPTER XVI. 1. What is the object of the chest? What are the lungs? 2. What their object, and how prepared for it? 3. Explain the air-vessels. 4. Describe the windpipe. The glottis, and the epiglottis. 5. What are the vocal chords? 6. How are they affected by disease? 7. Of what are the air-tubes composed?

L   14 *

CHAPTER XVII. 1. What is the mucous membrane? 2. What is its condition in health? What in a cold? 3. How is it affected by foreign matter — by a drop of water within the glottis? 4. How is coughing performed? 5. What other substances irritate these delicate organs? 6. Why is coughing excited? What causes a dry cough? 7. Describe the pulmonary arteries and their use. 8. The pulmonary veins. 9. The pulmonary circulation. Explain Fig. XI. 10. How much blood flows into and from the lungs a minute?

CHAPTER XVIII. 1. How is respiration performed? How inspiration? 2. How is the air forced out? 3. How often does this respiration take place? How much air is inhaled? 4. While the chest is expanding and contracting, what is the heart doing? 5. How are the blood and the air brought together? 6. What exchange takes place between them? 7. Of what does the waste matter mostly consist? What is carbon? 8. What is hydrogen? 9. What is atmospheric air? 10. What is oxygen? 11. Nitrogen? 12. What the relation between oxygen and carbon?

CHAPTER XIX. 1. How do the blood and air affect each other in the lungs? 2. What pass through the separating walls? 3. What effect has this change on the blood? 4. What on the air? 5. What part does the oxygen perform in this work? How is the air affected?

6. How much carbonic acid can the air take? How is suffocation sometimes caused? 7. What other limit has the air? 8. How much water does a healthy person exhale in a day? In what form? 9. In what way is the air spoiled for breathing?

10. How are several persons confined in a small, close room affected? 11. What are our sensations on going from the fresh air into a crowded room? 12. What is the condition of the air in these rooms? How does it affect the occupants? 13. What is the consequence, if the dead atoms are not carried away?

CHAPTER XX. 1. Upon what does the quantity of waste atoms depend? 2. Upon what do these changes depend? 3. How does exercise affect the removal of dead atoms? 4. What other circumstances affect it? 5. How does the state of the lungs affect it? 6. What is the effect of consumption, &c., and of compression of the chest? 7. What circumstances lessen the amount of air inspired, and what are the consequences?
8. To what does the size of the chest correspond, and why? 9. What is the intention of Nature in regard to the size of the chest? 10. Describe the natural chest. 11. What is the object of this arrangement, and how may it be abused? 12. What are the motions of the chest? What is necessary? What the effect of restriction? 13. Explain Fig. XII. 14. What are the objections to a tight bandage about the chest?

CHAPTER XXI. 1. Why do we need fresh air at every respiration? 2. How much fresh air does each individual need? 3. What is the difference between the air in and out of houses? 4. What is necessary for all inhabited rooms? 5. What is the effect of impure air?

ANIMAL HEAT.

CHAPTER XXII. 1. How much does the heat of our bodies vary? 2. What influence have surrounding objects on animal heat? Explain. 3. Whence is the animal heat? 4. What is the chemical process of burning wood? What is combustion? 5. How is the heat prepared in the animal body? 6. What are necessary for this internal fire? What effect has food? 7. What effect has air? 8. What effect has exercise? 9. How is it affected by difference in food? 10. Are our sensations sure tests of the degrees of heat and cold?

SKIN.

CHAPTER XXIII. 1. What is the object of the skin? 2. Of what is it composed? 3. Describe the cuticle. 4. To what laws is it subject? 5. How are its atoms changed? 6. What are the old particles? 7. How and where does the cuticle differ in thickness?

8. What is the effect of friction? 9. Where does the cuticle become thick by use? 10. 11. What is the effect of violent friction, and what example? 12. How are corns formed? 13. How removed? 14. Where is the seat of color? How affected?

CHAPTER XXIV. 1. Describe the true skin. 2. With what is this skin supplied? What effect has heat and cold on it?

3. What is the perspiratory apparatus? Explain Fig. XIII. 4. What office does this apparatus perform? 5. What are sweat and insensible perspiration? 6. What do you know of the amount of perspiration? 7. 8. With what does the amount vary? 9. When is it most free? 10. How is it affected by water-proof clothing?

11. What is the oily excretion? What is its object? 12. Of what use are these excretions?

13. What is the effect of cold on perspiration? 14. How do persons take cold? What the various effects?

15. What is the temperature of the body? How is heat passed off? 16. What is the most healthy temperature of rooms?

CHAPTER XXV. 1. What is cutaneous absorption? 2. What illustrates its action? 3. When is this most active? 4. When is a person most easily affected by poison or contagion? 5. How is cutaneous absorption affected by foulness of the skin? 6. Whose skin is most frequently diseased?

7. In which layer is the seat of sensation? 8. Where is the greatest sensibility? 9. Where is the sense of touch? Can it be improved? Example. 10. What difference between the draper and bricklayer?

CHAPTER XXVI. 1. Of what use is clothing? 2. On what must the quantity depend? 3. How does habit affect the need of clothing? 4. How does exposure?

5. What is the object of clothing, and how is it effected? 6. What texture of clothing, and what adaptation to the body, are warmest? 7. What is the character of wool? 8. What of cotton? 9. What of silk and linen? 10. Why is cotton better for warm climates than linen? 11. What becomes of the excretions of the skin? 12. Why should clothes be frequently changed? 13. What other way of partially purifying garments? 14. In what way, and how often, should this be done? 15. How should beds be treated?

CHAPTER XXVII. 1. What interrupt the functions of the skin, and how? 2. How often should these be removed? 3. What parts of the body need cleansing, and why? 4. What effect has bathing on the skin?
5. What effect has the cold bath? 6. How often should it be taken? 7. How can one gain power to endure it in cold weather? 8. When should the cold bath be omitted? 9. What beside the skin is benefited by this bath? 10. What are its results? 11. How should the stomach affect bathing? 12. When is the best time for a bath? Why? 13. Why should one who is cold not take a cold bath? 14. Why should a person warm himself before going into the cold air? 15. In what way is the sense of touch blunted? How improved?

BONES.

CHAPTER XXVIII. 1. What is the use of the bones? 2. Of what are they composed? 3. What mostly in early years? 4. What in old age? 5. What in the middle periods of life? 6. How are the bones of different ages affected by falling? 7. With what are the bones supplied, and to what law are they subject? 8. Have they sensibility?
9. What is the effect of exercise on the bones? 10. Explain Fig. XIV. 11. Of what is the head composed? 12. Of what the trunk? 13. The upper extremity? 14. The lower?

CHAPTER XXIX. 1. Why is the skull round, and what is its use? 2. What are the bones of the spinal column? Explain Fig. XV. and XVI. How are the vertebræ arranged? 3. What substance between these bones? 4. What its character and use? 5. What is the natural form of the spine?
6. What bones compose the chest? 7. Describe the wrist. Explain Fig. XVII. 8. What is the arrangement of the hand? Explain Fig. XVIII.
9. How are the bones of the foot arranged? Explain Fig. XIX. and XX. 10. Where on the foot do the bones of the leg rest? Why? 11. In walking, where is the weight of the body first received? Why no jar? 12. Why does a man feel a jar when he walks on a ladder?

CHAPTER XXX. 1. How are the bones connected? 2. Explain the hinge joint. 3. The ball and socket joint. 4. The joints of the spine. 5. How is the head attached to the back-bone? 6. Why are cartilages in the joints?
7. What is the synovial membrane? 8. How is it sometimes diseased? 9. How are the bones held together? 10. What is a dislocation? What a sprain? 11. What can you say of the structure and power of the human frame?

CHAPTER XXXI. 1. 2. 3. What advantage in the erect attitude? 4. What other advantages? Examples. 5. How is the strength affected by the attitude? 6. What is the effect of curvature of the spine? 7. How does it affect the whole body? 8. How is a stooping habit sometimes acquired? 9. How is the lateral curvature acquired?

MUSCLES.

CHAPTER XXXII. 1. What are the muscles? What power have they? 2. How do they bend the joints? 3. And what is their use? 4. What muscle bends the elbow? 5. What straightens it? What move the shoulder? 6. What move the fingers? Explain Fig. XXII. 7. What move the toes? 8. What is the form of the muscles? 9. How are the muscles arranged to perform the motions of the body? 10. What are the voluntary, and what the involuntary muscles?

CHAPTER XXXIII. 1. Explain the action of the voluntary muscles. 2. What muscles lay the hand on the head? 3. Explain the action of the muscles in walking. 4. What motions are produced by the muscles? Explain Fig. XXIV. 5. How are the muscles increased in size? Examples. 6. In what ways is exercise beneficial? 7. How does it affect digestion? 8. What is required of sedentary persons? 9. How are students and sedentary mechanics benefited by exercise? 10. Can all take the same amount of exercise? Why? 11. What general rule for exercise? 12. What employments afford the best exercise? 13. When is the system best prepared for exercise?

CHAPTER XXXIV. 1. How much action can the system sustain? 2. What is the effect of excess of labor? 3. How can strength be best maintained? 4. What is the effect of judicious labor? 5. What of excessive labor? 6. How should labor be varied? What employments favor this? 7. Is it safe to make violent exertions? 8. What should be the exercise of youth, and why?

9. What is the effect of day and night labor? 10. What workmen are soonest exhausted? 11. How does labor affect the demand for food? 12. How is the power for labor affected by the organs of digestion, circulation, and respiration? 13. How do diseased lungs prevent violent exercise? 14. How does tight clothing? 15. What is necessary for the strength of the muscles?

16. What effect has spirit on labor? 17. What have cheerfulness and melancholy?

CHAPTER XXXV. 1. What is the skull? 2. Describe the brain. 3. The coverings. 4. What vessels has it? To what laws is it subject? 5. What is the spinal cord? De-

www.ingramcontent.com/pod-product-compliance
Lightning Source LLC
Chambersburg PA
CBHW030246170426
43202CB00009B/641